Green Energy and Technology

Climate change, environmental impact and the limited natural resources urge scientific research and novel technical solutions. The monograph series Green Energy and Technology serves as a publishing platform for scientific and technological approaches to "green"—i.e. environmentally friendly and sustainable—technologies. While a focus lies on energy and power supply, it also covers "green" solutions in industrial engineering and engineering design. Green Energy and Technology addresses researchers, advanced students, technical consultants as well as decision makers in industries and politics. Hence, the level of presentation spans from instructional to highly technical.

Indexed in Scopus.

Indexed in Ei Compendex.

Ali Ahmadian • Ali Diabat
Ali Almansoori • Ali Elkamel
Editors

Sustainable Energy Resources in Smart Cities

Springer

Editors
Ali Ahmadian
Department of Electrical Engineering
University of Bonab
Bonab, Iran

Ali Almansoori
Department of Chemical Engineering
Khalifa University
Abu Dhabi, United Arab Emirates

Ali Diabat
Division of Engineering
New York University Abu Dhabi
Saadiyat Island, Abu Dhabi
United Arab Emirates

Department of Civil and Urban Engineering, Tandon School of Engineering
New York University
Brooklyn, NY, USA

Honorary Professor, School of Engineering
The University of Jordan
Amman, Jordan

Ali Elkamel
Research and Innovation Center on CO_2 and Hydrogen (RICH Center) and Chemical and Petroleum Engineering Department
Khalifa University of Science and Technology
Abu Dhabi, United Arab Emirates

Department of Chemical Engineering
University of Waterloo
Waterloo, ON, Canada

ISSN 1865-3529 ISSN 1865-3537 (electronic)
Green Energy and Technology
ISBN 978-3-031-91515-4 ISBN 978-3-031-91516-1 (eBook)
https://doi.org/10.1007/978-3-031-91516-1

This Springer imprint is published by the registered company Springer Nature Switzerland AG
The registered company address is: Gewerbestrasse 11, 6330 Cham, Switzerland

Preface

Sustainable energy resources play a crucial role in shaping the future of smart cities, driving the transition toward sustainable development. As urbanization accelerates, the demand for energy continues to rise, making it essential to adopt renewable and efficient energy solutions that minimize environmental impact. By integrating solar, wind, hydro, and other clean energy sources into urban infrastructure, smart cities can reduce carbon emissions, enhance energy security, and promote economic resilience. These advancements not only contribute to environmental preservation but also improve the quality of life for residents by fostering cleaner air, reducing energy costs, and enabling intelligent energy management systems. Ultimately, the widespread adoption of sustainable energy resources is fundamental to building self-sufficient, eco-friendly, and technologically advanced cities that align with global sustainability goals.

This book is an attempt to bring together experts from different disciplines related to sustainable energy resources in smart cities and its impact on sustainable development. It contains eight chapters in which numerous researchers and experts from academia and industries collaborated. The breakdown of the chapters is as follows:

- Chapter 1 delves into the intricate relationship between energy considerations and urban planning within the context of smart cities, explores the potential of renewable energy sources, and examines strategies for their effective implementation. Energy efficiency and the management of energy consumption in urban areas are crucial aspects to consider. Various components of cities can be transformed into smart entities, including smart grids and smart water grids. Additionally, investments in energy storage solutions are pivotal for the development of smart cities and their infrastructure. Combined heat and power (CHP) units represent a viable option for integration in urban settings, operating either through electricity and water or electricity and gas.
- Chapter 2 summarizes the work that has been conducted in the field of geothermal energy resources in Jordan. The need to explore alternative solutions through renewable energy sources has arisen due to the global fossil fuel crisis worldwide.

Renewable and sustainable energy sources including solar, wind, hydroelectric, biomass, and geothermal energy have attracted extensive attention of many researchers worldwide. Geothermal energy sources are considered inexhaustible, clean, resilient to seasonal fluctuations, and stable against the impacts of climate change. Therefore, it is important to investigate and study the potential of using geothermal energy sources for different applications, including space heating, cooling, recreational, desalination, and power generation applications.

- Chapter 3 discusses the evolution of energy efficiency regulations in the European Union (EU), focusing on Smart Readiness Indicator (SRI) derived from Energy Performance of Buildings Directive (EPBD) and Energy Efficiency Directive (EED).
- Chapter 4 presents optimal operation of integrated electricity and natural gas distribution networks in smart cities.
- Chapter 5 schedules a multi-energy networked microgrid in smart cities under transactive energy concept.
- Chapter 6 examines the integration of various energy forms, including photovoltaic solar power (PV), hydrogen fuel cells (FC), and battery capacity (BAT), into an on-grid hybrid energy system (HES). This study introduces a multi-objective optimization model to address the cost-emission challenges of a PV-FC-BAT HES, incorporating the effects of a demand response program (DRP).
- Chapter 7 provides a comprehensive review on microgrid and nanogrid definitions. Components of these grids as well as characteristics of each components are presented. Then, applications of microgrids and nanogrids in smart cities are classified and introduced. Microgrid and nanogrid design and related issues are also discussed. International standards and codes associated with microgrid and nanogrid design and operation are introduced focusing on aims, scopes, and key technical and environmental restrictions. Finally, recent projects developed in developed countries are introduced.
- Chapter 8 focusses on the implementation of smart homes and buildings within smart cities.

The editors of this book warmly thank all the contributors for their valuable works. Also, we would like to thank the respected reviewers who improved the quality of this book by their valuable and important comments.

Bonab, Iran — Ali Ahmadian
Abu Dhabi, UAE — Ali Diabat
Brooklyn, NY, USA
Amman, Jordan
Abu Dhabi, UAE — Ali Almansoori
Abu Dhabi, UAE — Ali Elkamel
Waterloo, ON, Canada

Contents

1 **An Overview of Sustainable Energy Resources in Smart Cities**..... 1
Armin Bagherian, Mohsen Gharehdaghi, Feria Shaker, Mohammad Mohsen Hayati, and Ebrahim Babaei

2 **A Review of Geothermal Energy Resources and Applications in Jordan: Current Status and Prospects**............ 45
Heba Alzaben

3 **Advancing Energy Efficiency in Smart Cities: The Smart Readiness Indicator (SRI) Framework for Sustainable Building Performance**.. 65
Armin Bagherian, Mohsen Gharehdaghi, Mohammad Mohsen Hayati, Farbod Borhani, Mehdi Abapour, and Behnam Mohammadi-ivatloo

4 **Optimal Operation of Integrated Electricity and Natural Gas Distribution Networks in Smart Cities** 97
Asma Nasiri, Nima Nasiri, Ebrahim Babaei, and Sajad Najafi Ravadanegh

5 **Multi-Energy Networked Microgrid Scheduling in Smart Cities Under Transactive Energy Concept** 127
Nima Nasiri, Asma Nasiri, Sajad Najafi Ravadanegh, Behnam Mohammadi-ivatloo, and Mehdi Abapour

6 **A Cost-Emission Trade-offs in Hybrid Energy Systems Using a Multi-Objective Approach with Demand Response Integration**... 151
Mehran Ghahramani, Mehrdad Ghahramani, and Mehdi Abapour

7 Microgrid and Nanogrid Implementation in Smart Cities 175
Omid Shahhoseini

8 Smart Home and Building Implementation in Smart Cities 203
Amirreza Samadi Bonab and Parisa Shirinabadi

Index 223

About the Authors

Ali Ahmadian is a Visiting Professor with the Department of Chemical Engineering at the University of Waterloo, Canada, and an Associate Professor with the Department of Electrical Engineering at the University of Bonab, Iran. He holds a Ph.D. in Electrical Engineering with a strong background in power and energy systems analysis. He has published over 100 papers in journals and conference proceedings and has received several national and international research awards. His research interests include transportation electrification, energy and environment, energy economics, and smart grid.

Ali Diabat is a Global Network Professor of Civil and Urban Engineering at New York University Abu Dhabi. Prior to this, he was a Full Professor of Engineering Systems and Management at Masdar Institute of Science and Technology in Abu Dhabi. He received his B.Sc. degree in Mechanical Engineering from Jordan University of Science and Technology, his M.Sc. degree in Operations Research from North Carolina State University, and a Ph.D. in Industrial Engineering from Purdue University in 1999, 2003, and 2008, respectively. Dr. Diabat's research focuses on different applications of optimization and operations research, in particular, logistics and supply chain management, healthcare logistics optimization, and production planning. Dr. Diabat has published over 150 research journal papers (75 of them since he joined NYUAD in June 2017) and over 30 conference papers in leading journals and international conference proceedings. Many of his publications are co-authored with his graduate students and, in many cases, with national and international collaborators. Dr. Diabat's research has received about 5 million dollars of grant funding from different industries and collaborative proposals with academic institutions. His industry experience includes working in the oil shipping industry in the Hashemite Kingdom of Jordan. After that, he held a position as a business analyst at United Airlines and two positions as an operations research analyst in the banking and beverages industries, all in the United States. Dr. Diabat has received several teaching awards, including the Outstanding Graduate Instructor Award and the Excellence in Teaching Award from Purdue University in 2004 and 2006, respectively. Dr. Diabat was the recipient of the 2012 Excellence in Teaching

Award from Masdar Institute, an achievement of special significance as it was the first award of its kind to be offered by the institute. In 2014, he also received the Best Faculty Research Award from the Department of Engineering Systems and Management at Masdar Institute. The most recent award was the Best Engineering Teaching Award in 2022. Dr. Diabat is an Associate Editor of the *SME Journal of Manufacturing Systems* (*JMS*) and an Area Editor of the *Journal of Computers and Industrial Engineering.* He is a member of the Mohammed Bin Rashid Academy of Scientists (MBRAS), a platform for the science community within the UAE that includes a selection of scientists and experts foreseeing the future of developments in science and technology.

Ali Almansoori is a Professor of Chemical Engineering and Associate Provost for Education at Khalifa University in Abu Dhabi. He holds a B.Sc. in Chemical Engineering with the highest distinction from the Florida Institute of Technology, a Ph.D. in Chemical Engineering in Process Systems Engineering from Imperial College London, and an Executive MBA from the London Business School. His specific research interests are in computer-aided modeling, optimization, and simulation, with applications in energy system design, sustainable operations, and supply chain management. He has published over 90 journal articles (of which one appeared in *Science* magazine), co-authored seven book chapters, presented more than 60 conference papers in local and international conferences or workshops, and supervised over 20 graduate students. He was the principal investigator for more than ten projects sponsored by industry and academia. He is a theme lead at KU Research Innovation Center on CO_2 and H_2 (RICH) in H_2 Energy Systems. He has received ten awards for his research contributions and impact on the scientific community, including the Mohammed Bin Rashid Medal for Scientific Excellence in 2019 and the 2021 Khalifa Award for Education in Higher Education. In addition to his academic experience, Prof. Almansoori has 8 years of administrative experience in higher education. He was the Chair of the Department of Chemical Engineering, the Dean of the College of Engineering, and the Interim Senior Vice President for Academic Affairs. He also managed the portfolio of policy and performance in higher education at the Executive Council of Abu Dhabi during his secondment in 2013.

Ali Elkamel is a Professor and Chair of Chemical and Petroleum Engineering Department at Khalifa University, Research and Innovation Center on CO_2 and Hydrogen (RICH) theme lead, and an Emeritus Professor at the University of Waterloo, where he is also cross-appointed in Systems Design Engineering and Management Science Department. Professor Elkamel holds a B.Sc. in Chemical Engineering and a B.Sc. in Mathematics from the Colorado School of Mines, an M.S. in Chemical Engineering from the University of Colorado Boulder, and a Ph.D. in Chemical Engineering from Purdue University. His specific research interests lie in computer-aided modeling, optimization, and simulation, with applications in energy production planning, carbon management, sustainable operations, and product design. He is currently focusing on research projects related to gas production and processing, the integration of renewable energy into oil and gas operations, and using data analytics (digitalization), machine learning, and artificial intelligence (AI) to improve the process and enterprise-wide efficiency and profitability. Prof. Elkamel's activities include supervising postdoctorate and research associates, advising graduate and undergraduate students, and participating in university and professional societal activities. He is also engaged in initiating and leading academic and industrial teams and establishing international and regional research collaboration programs with industrial partners, national laboratories, and international research institutes. Prof. Elkamel influence extends beyond the classroom, as he has supervised over 75 Master's students, 45 PhD students (20 in academia) and more than 45 postdoctoral fellows/ research associates. Notably, all of his trainees have successfully secured positions in the chemical process industry and academia. He has been funded for several research projects by the government and industry. He has more than 450 journal articles, 185 proceedings, and 50 book chapters and has been an invited speaker numerous times at academic institutions worldwide and national and international conferences. He has co-authored seven books, including *Planning of Refinery and Petrochemical Operations and Environmentally Conscious Fossil Energy Production.*

Chapter 1
An Overview of Sustainable Energy Resources in Smart Cities

Armin Bagherian, Mohsen Gharehdaghi, Feria Shaker, Mohammad Mohsen Hayati, and Ebrahim Babaei

1.1 Introduction

The significant importance of sustainable energy sources in smart urban areas is crucial for enhancing a more effective and less polluted environment [1]. With the ongoing global urbanization trend, smart cities are increasingly emphasizing the incorporation of sustainable energy sources to address their expanding energy requirements [2]. This study aims to explore diverse sustainable energy sources deployed in smart cities, encompassing their advantages, obstacles, and sophisticated technologies employed to enhance their effectiveness. Comprehending the array of sustainable energy sources in smart cities offers valuable perspectives for cultivating eco-friendly and resilient urban settings. Solar power, wind power, and geothermal energy are sustainable energy sources that are assimilated into smart cities [3]. Solar power and wind power represent sustainable energy sources that provide benefits such as reducing greenhouse gas emissions and decreasing the reliance on fossil fuels [4]. Solar power, an eco-friendly energy option that captures sunlight for electricity generation, stands as a clean and long-lasting energy solution that can be integrated into smart city infrastructures to mitigate their environmental impact [5]. Wind power, another sustainable energy alternative, utilizes the wind's energy for electricity production [6]. It serves as a clean and abundant energy choice that can be employed in smart cities to lower their carbon footprint and reduce their dependency on conventional energy sources [7]. Geothermal energy, a sustainable energy form that taps into the earth's

A. Bagherian · M. Gharehdaghi · F. Shaker · E. Babaei (✉)
Faculty of Electrical and Computer Engineering, University of Tabriz, Tabriz, Iran
e-mail: arminbagherian@ieee.org

M. M. Hayati
Faculty of Electrical and Computer Engineering, Energy Systems Research Institute (ESRI), University of Tabriz, Tabriz, Iran
e-mail: mohammadmohsen.hayati@tabrizu.ac.ir

A. Ahmadian et al. (eds.), *Sustainable Energy Resources in Smart Cities*, Green Energy and Technology, https://doi.org/10.1007/978-3-031-91516-1_1

natural heat for power generation and heating purposes, emerges as a reliable and enduring energy option that can provide a stable power source for smart cities, thereby diminishing their reliance on traditional energy supplies [8–10].

Integrating sustainable energy sources into smart cities offers the advantage of decreasing air pollution and enhancing air quality, thus fostering a healthier and more sustainable urban habitat [11]. Diversifying energy sources away from fossil fuels can bolster energy security and fortitude in smart cities, guaranteeing a steady energy provision even amidst possible disturbances [12, 13]. This shift can fortify the energy framework of smart cities, diminishing the likelihood of blackouts and elevating resilience against severe weather occurrences [14].

The incorporation of sustainable energy sources in smart urban areas can also generate employment opportunities, given that the establishment and upkeep of renewable energy infrastructure can spawn fresh job prospects and foster economic advancement within the locality [15, 16]. This could yield a livelier and more sustainable local economy, along with the possibility of novel business prospects in the renewable energy domain [17]. By integrating sustainable energy sources, smart cities can additionally curtail their overall energy expenditures and diminish their dependence on costly and unpredictable fossil fuels, resulting in enduring financial advantages [18]. The utilization of sustainable energy sources further aids in lowering greenhouse gas emissions, aiding in the fight against climate change and alleviating its environmental repercussions, thereby fostering a more resilient and sustainable urban setting [19, 20].

The incorporation of renewable energy sources (RES) in smart urban areas can stimulate creativity and technological progress, resulting in the enhancement of more productive and efficient energy solutions for the urban environment [21]. This advancement can facilitate the establishment of a more sophisticated and contemporary energy framework, enhancing the general standard and dependability of energy provision in smart cities [22]. The assimilation of sustainable energy sources in smart cities can additionally aid in broadening the energy blend, lessening the city's reliance on a sole energy origin, and heightening overall energy autonomy and security [23]. This progression can pave the way for a more enduring and robust energy structure for the urban setting [24].

Energy efficiency is essential for tackling the issues related to worldwide climate change, as well as improving energy supply reliability, business competitiveness, and societal welfare [25]. Considering electrification to achieve energy efficiency is crucial for promoting energy preservation [26]. There are numerous opportunities to improve energy efficiency in buildings, transportation, and industries, which can result in energy conservation by shifting from inefficient fossil fuel systems to more efficient electrical technologies. Such shifts bring benefits in terms of finances, the environment, health, and assets [27, 28].

Advanced metering infrastructure (AMI) fills this gap through linking the power generation grid and consumers by bidirectional exchange of information. Applying smart meters enables the users to monitor the exact and real-time electricity usage by affording dynamic price information, and their contribution on the carbon emissions, and assisting them to make more intelligent decisions [29, 30].

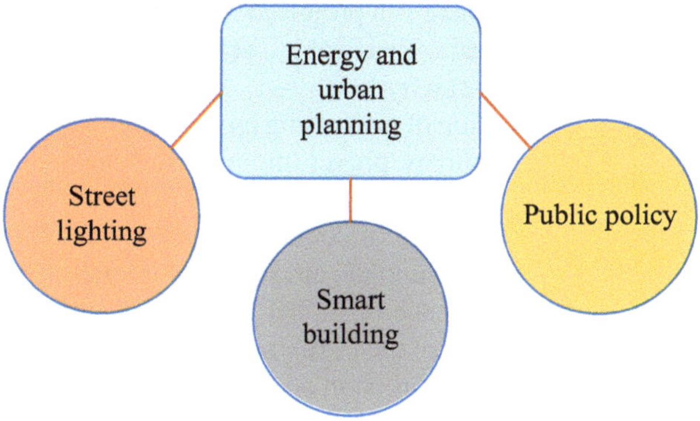

Fig. 1.1. Key focus in urban planning and technological progress [2]

1.2 Energy and Urban Planning

Urban design relies on the collaboration of various entities such as governmental bodies, public institutions, NGOs, businesses, and communities to achieve energy sustainability. This collaboration is essential for enhancing energy security, maintaining resource availability, and promoting sustainable production. Emphasizing the establishment of effective and sustainable public lighting systems and the creation of intelligent residences, structures, and communities that promote energy efficiency and the adoption of renewable energy sources should be a key focus in urban planning. Technological progress, like automating lighting and addressing thermal comfort (shown in Fig. 1.1), can cut energy consumption and boost effectiveness. Retrofitting offers a budget-friendly way to blend new ideas and eco-friendliness into older structures. To better urban sustainability and enhance life quality in cities worldwide, embracing a comprehensive strategy covering energy, eco-friendliness, and intelligent urban areas is crucial [2].

1.3 Renewable Energy Sources

Renewable energy harnesses natural processes to tap into sources like solar radiation, wind movement, water reservoirs stored energy, geothermal heat from the Earth's core, and bioenergy from plants. Various technologies are employed to convert these abundant resources into sustainable energy forms, predominantly electricity, with additional uses in heat generation, chemical reactions, and mechanical power generation.

Utilizing coal, oil, and natural gas as the primary energy sources for residential heating, transportation, and fuel production is convenient; nevertheless, it is crucial to acknowledge that these resources are limited on Earth. The consumption rate surpasses the production rate significantly, leading to eventual depletion of these finite resources. Concurrently, there is a projected 33% surge in the nation's energy needs over the upcoming 20 years. Renewable energy stands as a promising solution to mitigate the impending scarcity [31].

Despite the theoretical situation of possessing an unlimited supply of fossil fuels, selecting renewable energy sources proves more beneficial for the ecosystem. Renewable energy setups are occasionally labeled as "clean" or "green" because of their limited or nonexistent discharge of contaminants. Conversely, the burning of fossil fuels emits greenhouse gases into the air, which subsequently trap solar radiation and contribute to the occurrence of global warming. Climate experts widely agree that the average temperature of the Earth has risen over the past century. If this trend continues, there will be an elevation in ocean levels, resulting in a higher probability of inundations, extreme heat events, droughts, and other intense weather patterns, as forecasted by researchers.

During the burning of fossil fuels, extra pollutants are released into the air, soil, and water. These pollutants greatly affect the environment and people. Respiratory ailments like asthma worsen due to air pollution. Acid rain, a consequence of sulfur dioxide and nitrogen oxides emissions, harms plants and fish. Nitrogen oxides additionally play a role in smog formation [32].

Some renewable energy sources that can be used in smart cities are shown in Fig. 1.2.

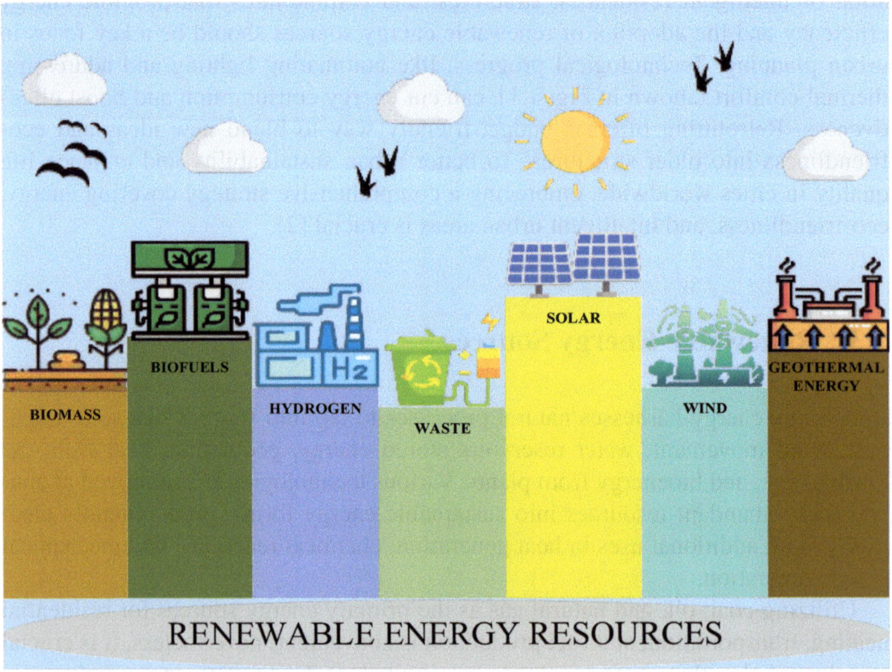

Fig. 1.2 Renewable energy resources in smart cities [33]

1.3.1 Solar Energy

Solar energy is a highly promising alternative energy source that can provide energy for all secondary sources, except for geothermal energy. The annual solar flux that reaches the Earth's atmosphere is immense, measuring approximately 5.661024 J. This amount is 5000 times greater than the total energy needs of humans for an entire year. Nevertheless, 35% of this energy is reflected by the Earth's surface, while the remaining portion is utilized for heating the Earth's atmosphere, the evaporation and condensation cycle in the atmosphere, the development of waves in seas and oceans, air and ocean currents, wind, and photosynthesis [34].

Solar energy engineering is an emerging sector of business that experiences tremendous growth, with a global yearly capital investment of $415 billion. In 2014, solar energy collectors in Europe produced about 100 million MWh, while global solar-radiation-based energy generation amounted to 186 million MWh. In 2014, the combined capacity of photoelectric installations amounted to 179 gigawatts (GW) [34].

Two main approaches exist for utilizing solar energy in generating power: photovoltaics (PVs), which employ photoelectric converters and solar cells, and thermal solar power generation.

In the 1980s, scientists pioneered the original thin-film solar cell employing economical noncrystalline silicon, igniting a surge in solar power exploration [35]. The expense of producing electricity through photovoltaic cells has decreased by roughly 30-fold over five decades, mainly attributed to the diminished expense of photovoltaic modules. Annually, fresh approaches surface to curtail economic outlays in this sector. Among these strategies is the integration of heterojunctions, particularly involving gallium and aluminum arsenide. Furthermore, the utilization of solar concentrators with concentration ratios spanning from 50 to 100 is under investigation.

Solar power is a renewable energy source suitable for generating electricity in various types of buildings. Solar power plants consist of modules with a parabolic concentrator for capturing solar energy and a receiver for heating the working fluid. Compared to towers, these modules are more economical at moderate capacities. While thermal solar power plants can be integrated into conventional thermal systems, their competitiveness in local markets is hindered by their high costs. Solar energy can be transformed into direct electrical current by utilizing solar-cell arrays, which are composed of thin layers of silicon or another semiconductor material. Photovoltaic converters (PVC) provide benefits such as exceptional dependability, steadfastness, and extended lifespan. Nevertheless, they exhibit exorbitant expenses and suboptimal effectiveness.

Solar-cell arrays are utilized for energy provision in outer space, whereas on Earth, their primary applications include powering remote consumers, radio location, and low-duty electronic facilities. The Ivanpah electric power station, located in the Mojave Desert, California, is the most extensive endeavor in this industry. The system consists of 300,000 controlled flat mirrors that focus sun radiation onto

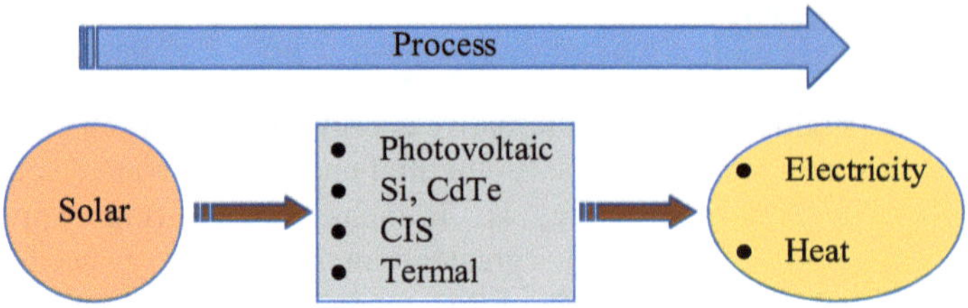

Fig. 1.3 The process of solar energy generation in smart city [37]

three identical 140 m high towers. Solar radiation's heat transforms water into steam, which is then directed towards turbine blades to generate electrical energy.

Nevertheless, the field of solar energy engineering has significant environmental challenges, including elevated air temperature and the possibility of water vapor condensation. Large-scale solar power facilities necessitate huge geographical expanses and should exclusively be constructed in arid regions. In addition, the production of mirrors and materials for solar cells involves dangerous chemical procedures, and the issue of proper disposal of these products must be resolved. To summarize, the production of solar energy encounters notable obstacles such as substantial initial investments, intricate machinery, fluctuating conditions, and uncertain energy generation [36]. Figure 1.3 illustrates the process of solar energy generation in smart cities.

1.3.2 Wind Energy

The future sustainability of wind energy production is limited due to the exhaustion of the most favorable places for wind energy installations. The capacity of a wind generator is determined by the total area covered by its blades and the height above the ground. Offshore sites near the coast are the most suitable locations for deploying wind generators. However, the cost of constructing them is 1.5 ± 2 times greater compared to onshore development.

Wind energy is a costly and eco-friendly substitute for fossil fuels, yet it is not now economically viable in the majority of markets. Building wind power facilities necessitates substantial land expanses, while the production of aluminum and/or glass-reinforced polymers poses risks. In the immediate future, it will be imperative to address the issue of disposing older or obsolete equipment. An additional significant issue associated with wind generators is the pronounced vibration of the supporting components, which is transmitted to the ground. The presence of noise generated by a wind power plant and its surrounding locations has detrimental impacts on both humans and many animal species, rendering them unsuitable for habitation. Wind turbines can cause disruptions to TV broadcasts and radio frequencies, and installations in areas with strong winds can disrupt the migration patterns of birds and insects. The irreversible impacts of widespread wind energy generation

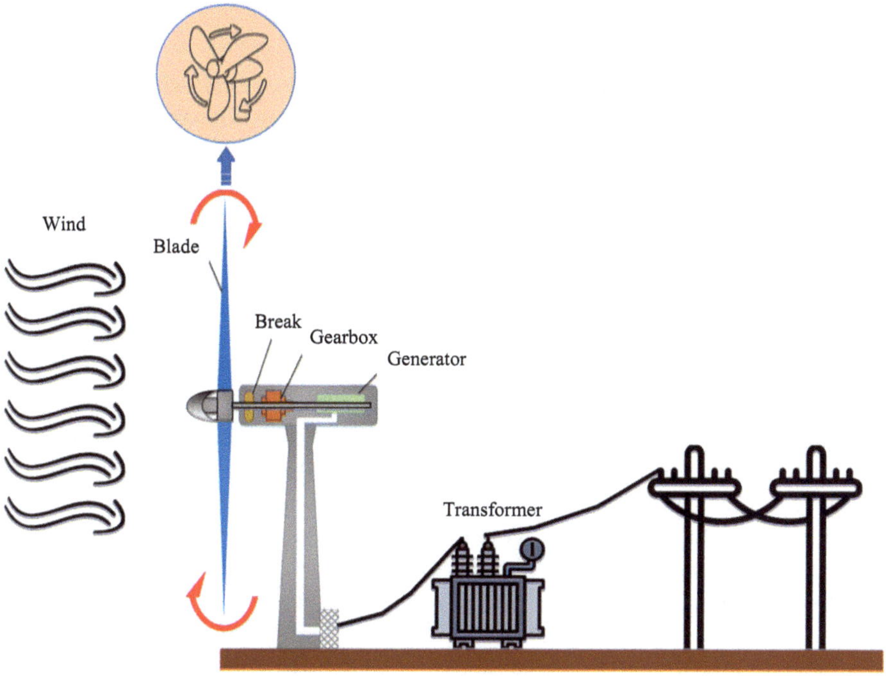

Fig. 1.4 Process of wind power generation flow in smart city

encompass the dispersion of air currents, diminished wind patterns, and disturbances in climate as well as the transmission of heat and moisture. Figure 1.4 depicts the flow of wind power generation in smart cities [36].

1.3.3 Hydroelectric Energy

The substantial cost-effectiveness of large hydropower plants (HPP) is attributed to their extended construction durations. Obstacles related to the environment and society impede the exploitation of the untapped hydropower capacity. The growth in water requirements may impede the progress of hydropower facilities and diminish the water capacity accessible for current installations. The creation of man-made reservoirs during dam erection leads to alterations in climate, disturbances in ecosystems, and a rise in soil salt levels. Furthermore, the establishment of water storage areas disrupts both ecosystems and the natural habitats for fish reproduction.

Upstream elevation of water at HPP facilities causes the formation of marshes, heightened soil salt content, and alterations in plant life and local climate. To illustrate, the Volga River has transformed into a series of shallow reservoirs, leading to a depletion in the river's energy capacity and the extinction of the Russian sturgeon [38].

1.3.4 Geothermal Energy

Geothermal energy engineering harnesses the heat energy stored within the Earth's interior to generate power. Within volcanic regions, water flowing through underground layers at shallow depths undergoes superheating, beyond the boiling point. This heated water then ascends to the surface through fractures and erupts as geysers. Well drilling provides access to the subterranean hot water. More commonly found are arid high-temperature rocks, which can be utilized for energy generation through the process of injecting water and extracting superheated water. Geothermal energy is extensively utilized in numerous countries due to the presence of high rock beds with temperatures ranging from 5 to 100 °C, particularly in geologically inactive areas.

Geothermal energy can be categorized into two types: Petro thermal and hydrothermal energy. Petro thermal energy is directly linked to the Earth's subsurface temperature, which increases by an average of 2.5 °C for every 100 m of depth. To harness energy, a pair of wells are bored. Water is injected into one of the wells, and as it migrates to the neighboring well, it is heated and transformed into steam. The primary issue linked to this form of energy generation is its lack of cost-effectiveness [39].

1.3.5 Biomass Energy

Biomass energy and solar energy are the primary means by which humanity's energy demand can be met through renewable sources. Combustion of biomass, primarily firewood and agricultural wastes, fulfills a substantial amount of household energy needs in developing nations. However, given the worldwide food shortage and deterioration of agricultural land, it is not feasible to meet the fast-increasing energy demand by sacrificing green energy.

Various technologies exist for utilizing bio-based feedstock, spanning across multiple iterations of biofuels. The initial iteration involves traditional agricultural produce converted into biodiesel and ethanol. Conversely, the subsequent iteration comprises inedible remnants from cultivated crops, grasses, and unused timber. Algae, categorized as a bio-based feedstock of the third generation, can be grown without land requirements, capable of generating substantial biomass quickly. However, harnessing natural water reservoirs may pose environmental risks due to the introduction of artificially engineered species into the ecosystem.

Phytoplankton production in artificial water reservoirs along the shoreline is seen as a promising advancement in numerous European countries. The latest feedstock technology employs methane fermentation and biomass hydroxylation to create methanol as a renewable fuel source. This biological system presents notable economic advantages in contrast to alternative solar energy conversion techniques. Nevertheless, its advancement is impeded by the prevailing low oil prices. Biofuels can manifest in solid, liquid, or gas states and can be produced through pyrolysis, a method that converts biomass into a liquid form. Pyrolysis, which encompasses the breakdown of waste materials and residual biomass, holds the potential to cater to up to 20% of Germany's automotive fuel consumption. The production expense for this fuel is projected at 0.80 euros/l. The yield of fuel components derived from one metric ton of discarded wood amounts to 100 kg [40].

Biogas, which is produced through the process of fermenting organic waste (biomass), is a commonly utilized kind of energy. Biomass undergoes decomposition through the activity of methanogenic bacteria and can be derived from poultry manure, cattle farm waste, and domestic garbage. Biogas is extensively utilized for household uses in areas that have limited energy resources. Several European countries, including Sweden and Germany, are utilizing agricultural and wood processing waste to produce biogas, which is then used to generate energy and produce synthetic motor fuels. In 2008, China had 30 million individual biogas production units, making it the global frontrunner in utilizing this type of fuel. These units supplied fuel to 22% of the rural population [41].

Bioethanol is advantageous in reducing air pollution from motor vehicles due to its high oxygen content, which promotes more thorough combustion of fuel. Brazil, a country known for its manufacturing of bioethanol derived from sugar cane, has held the position of global leader in both bioethanol production and its utilization as a fuel for automobiles. Additional viable biofuels include biobutanol, which lacks corrosiveness and can be conveniently delivered using the current gasoline infrastructure.

Biofuels, originating from biomass and discarded resources, are increasingly favored as a sustainable energy alternative. Within Europe, biodiesel stands out as the primary biofuel, sourced from materials like rapeseed, soybean, palm, and coconut oils, along with food industry by-products and cooking waste. In the year 2010, the worldwide liquid biofuel production reached 105 billion liters, equivalent to 2.7% of the overall gasoline consumption in road transportation globally. Specifically, bioethanol production accounted for 86 billion liters, while biodiesel production amounted to 19 billion liters.

Currently, the top five producers of biofuels are actively engaged in exploring biomass utilization techniques, achieving notable progress in manufacturing bioethanol from farm residues and various algae species. However, the widespread utilization of lignin remains challenging. The idea of mitigating anthropogenic impacts on atmospheric activities through the expansion of biomass energy and the use of bio-derived raw materials is attractive.

The worldwide biofuel industry has reached a considerable magnitude, exceeding $100 billion. Russia stands out as a primary supplier of fuel pellets to Europe, utilizing merely 20% of its biofuel output domestically. The country possesses the capability to manufacture approximately 72 billion cubic meters of biogas each year, enabling the generation of 151,200 gigawatt-hours of electricity and 169,344 gigawatt-hours of heat.

Nevertheless, the utilization of biomass on a large industrial level leads to detrimental environmental effects, including soil deterioration and the eradication of

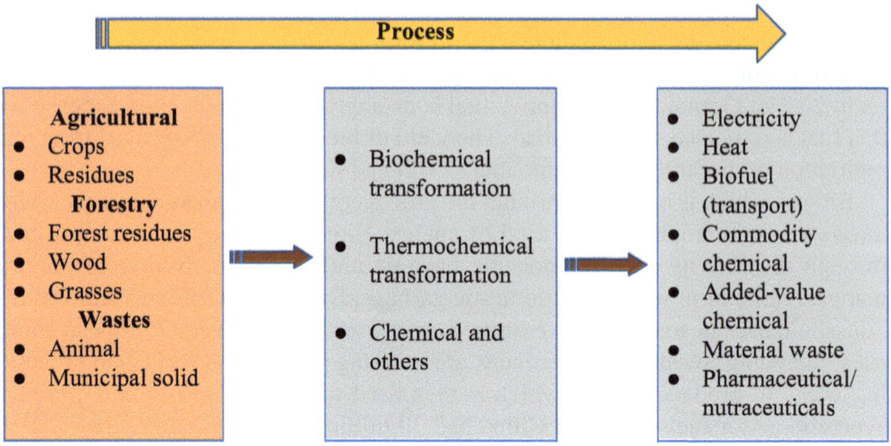

Fig. 1.5 Process of biomass conversion technology in smart city [37]

animal habitats and microecosystems. In addition, rapidly growing plants such as poplar, willow, and eucalyptus produce isoprenoids, which can react with atmospheric oxygen to form detrimental tropospheric ozone. Figure 1.5 depicts the process of biomass conversion technology in smart cities [37].

1.3.6 Hydrogen Energy

Hydrogen power is a widely embraced substitute energy option, backed by numerous nations implementing government initiatives for extensive industrial manufacturing and infrastructure to facilitate its use. When hydrogen is burned, it generates water, and its energy content surpasses that of hydrocarbon fuels significantly. Nevertheless, hydrogen serves as a secondary energy transporter and necessitates sourcing from plentiful and cost-effective raw resources. Presently, industrial-scale hydrogen generation relies on transforming fossil hydrocarbons such as natural gas, coal, and oil. The expense associated with producing hydrogen through alternative means like water electrolysis is notably higher, thereby constraining its widespread industrial utilization [42, 43].

The advancement of hydrogen energy generation from natural gas and other hydrocarbons is not very rational due to its generation of substantial quantities of carbon dioxide. A widespread transition to hydrogen fuel derived from natural gas reforming or water electrolysis would notably escalate carbon dioxide discharges, rendering the global environmental impacts of hydrogen energy generation nonsensical [44]. In order to viably generate hydrogen from water, currently, only three techniques show promise: biotechnology and microbiology, utilizing direct solar electricity for water electrolysis, and advancing photocatalysts that split water upon sunlight exposure. These strategies are in their initial phases and come with their own potentials and constraints [45].

1.3.7 *Waste Energy*

Biomass, originating from forestry and agricultural leftovers, municipal waste, and biogas, represents an additional type of sustainable energy (RE) that can be utilized in combined heat and power (CHP) systems for generating electricity and heat. Nevertheless, the incineration of biomass discharges carbon dioxide retained during plant development, underscoring the importance of integrating it into extensive energy production to prevent the release of greenhouse gases.

Urban biomass presents opportunities for comprehensive evaluations throughout the process of converting and using biomass as biofuels. Varieties of poplar with short growing seasons cultivated on underutilized lands in Boston, Massachusetts, have the potential to satisfy around 0.6% of the city's yearly primary energy requirements. Urban crops may be more suitable for nutritional purposes than energy production, serving as a local source of fresh produce for urban populations [46].

Hydrogen represents a prospective clean fuel option for energy storage and transportation in the future, owing to its independence from specific sources and its high energy density per unit mass compared to conventional fuel sources derived from petroleum. The processes of electrolysis and gas storage play key roles in the generation and transformation of hydrogen. Nevertheless, specialized tanks become necessary for storing hydrogen, as the gas is compressed to pressures ranging from 200 to 350 bar, which are suitable for domestic applications.

While waste to energy is not classified as entirely renewable, it could present advantageous opportunities for advancing sustainable energy strategies in urban areas. Urban dwellers generate nearly double the volume of waste compared to their rural counterparts, with waste-to-energy plants and landfill-gas-to-energy systems being prevalent methods of waste-to-energy conversion.

Direct combustion of urban solid waste in a waste-to-energy facility commonly produces around 0.6–1.8 megawatt-hours per metric ton of urban solid waste. Implementing combined heat and power generation in these facilities could utilize the surplus heat energy for warming purposes.

While there has been notable economic advancement and cost equivalency reached in comparison to traditional fossil fuels, renewable energy sources have struggled to demonstrate their competitiveness because of their reduced power concentrations, hindering their broad utilization in city settings [33]. Figure 1.6 illustrates a typical process for converting biomass into alternative energy.

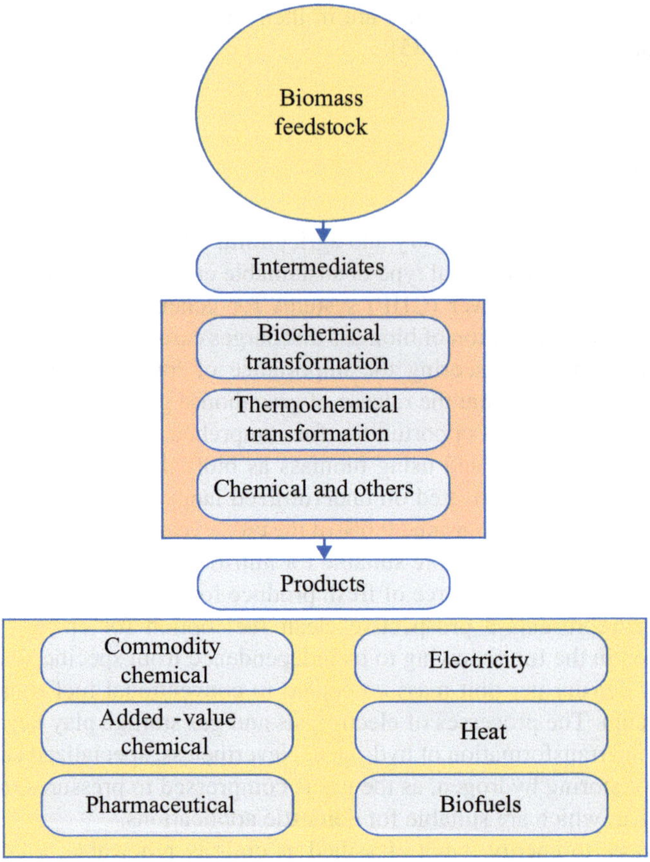

Fig. 1.6 A typical biomass conversion process into alternative energy [33]

1.4 Energy Efficiency Measures

Energy efficiency plays a vital role in urban settings as energy usage and environmental pollution continue to rise. The transition of populations from rural to urban regions poses sustainability challenges, highlighting the significance of energy efficiency in mitigating pollution and managing heat generated by various activities. The development of smart cities in the future holds promise in tackling environmental concerns and optimizing energy consumption.

Smart urban areas have the potential to tackle the challenges posed by population expansion and urbanization by enhancing the efficiency of resource utilization and cutting down on consumption levels through enhancements in infrastructure and alterations in behavior. Crucial components encompass the intelligent grid, structures, transit systems, amenities, infrastructure, monitoring systems, urban planning, and architectural design. Vital facilities such as power supply, gas provision, potable water distribution, transportation networks, communication systems, and

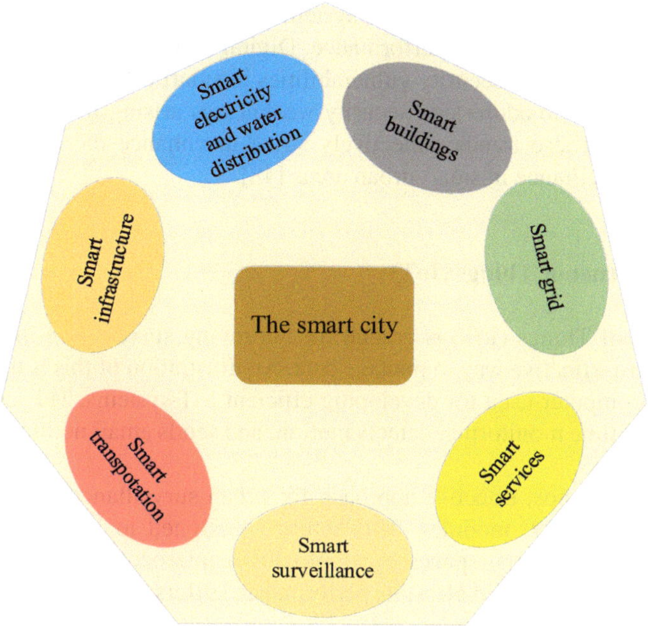

Fig. 1.7 Some items of active approach for development of smart cities [47]

sustainable energy sources ought to be unified within an automated framework. Administrative hubs can manage and regulate these strata utilizing decentralized sensor mechanisms and control technologies, which are classified as proactive methodologies that adapt dynamically to the city's requirements [47]. Figure 1.7 illustrates some examples of an active approach to the development of smart cities.

1.4.1 Smart Surveillance

Smart cities are turning to surveillance systems as their sensory tools. While conventional surveillance setups require human intervention for video analysis, upcoming smart cities will handle extensive data generation and processing. By employing smart devices such as vehicles, traffic signals, and mobile phones, advanced surveillance systems can enhance urban area supervision effectively. These setups play roles in safeguarding, guiding, organizing, and impacting individuals and assets. However, obstacles like data accuracy, privacy concerns, governmental backing, security measures, and the computational model persist.

The primary emphasis of research trends in smart urban areas lies in advancing surveillance systems through Internet of Things (IoT) technology. Within this domain, computer vision plays a crucial role in enhancing algorithms for object identification and monitoring in smart urban areas. Computational aspects are geared towards attaining surveillance systems that are cost-effective, require minimal storage, and deliver high performance. Digital image and video processing are instrumental in tackling security vulnerabilities in smart urban areas. The final category is dedicated to addressing security weaknesses arising from diverse surveillance systems. These trends collectively strive to enhance the effectiveness and security of surveillance in smart urban areas [48].

1.4.1.1 Internet of Things (IoT)

The Internet of Things (IoT) is crucial for improving surveillance in smart cities, offering a cost-effective way to manage data. An illustration of this is the Raspberry Pi, a small computer used for developing efficient IoT systems. This configuration provides real-time monitoring, detects motion, and sends email notifications for any suspicious events.

An alternative cost-effective approach for urban surveillance involves utilizing Arduino and CC2500 modules. This setup is designed to identify questionable movements in residential spaces by employing a microcontroller board, Arduino UNO, and CC2500. Nonetheless, its performance suffers when there is no access to the Internet.

An alternative approach is centered on minimizing data requirements in urban surveillance setups through the introduction of a straightforward algorithm for identifying vehicles and/or vehicle licenses. This technique is uncomplicated, feasible to deploy, and effectively captures crucial details from images, thereby cutting down on stored data.

Additionally, a surveillance system for traffic management employing ontology is suggested to forecast incidents and bolster security in intelligent urban areas. By utilizing a dynamic Bayesian network, the system calculates the likelihood of an incident occurring, thereby averting or mitigating potential harm. Another traffic surveillance system tailored for IoT-enabled intelligent roadways utilizes sensors installed along roadsides to oversee and identify vehicles, pinpoint road congestions, and alert users through an Android application [49].

1.4.1.2 Computer Vision

The idea of "smart cities" foresees forthcoming urban hubs as secure, eco-friendly, and effective by utilizing cutting-edge materials, sensors, electronics, and computerized setups. Research hurdles encompass a range of scientific fields such as physics, chemistry, biology, mathematics, and engineering. Essential are components with autonomous monitoring features, driving the need for system planning and incorporation to enhance resource management, preemptive maintenance, and safety. Sophisticated surveillance and intelligent sensors facilitate instant data gathering to support well-informed decision-making. The ultimate goal involves structures capable of self-repair and supervision of environmental conditions to promote a hygienic, safe, and efficient living space. Brookhaven National Laboratory spearheads an effort that harmonizes technology with urban obstacles to shape future city frameworks [50].

1.4.1.3 Computing

The advent of cloud computing has had a profound effect on smart cities, as it has facilitated the handling of large volumes of data and the integration of the Internet of Things (IoT). Deep neural networks (DNN), shallow neural networks (SNN), and fog nodes have been used to enhance surveillance systems in smart cities. These strategies are highly effective in terms of low latency, low cost, and quick decision-making. An advanced method based on trajectory analysis has been created to identify abnormal behavior in movies, offering reliable evaluation of vehicle density and event information in real time. The system utilizes the MQTT protocol to distribute video summarization, ensuring scalability and cost-effectiveness. Nevertheless, the system's memory usage is substantial as a result of employing Apache Spark and Hadoop, compounded by the Raspberry Pi's constrained RAM capacity. Cloud computing is crucial for optimizing monitoring systems in smart cities [51].

1.4.1.4 Digital Image and Video Processing

The smart city system offers a holistic approach to economic and social progress, urban infrastructure enhancement, citizen welfare, and environmental preservation. It integrates a video monitoring setup comprising a front-facing camera, transmission wiring, and control mechanism. These cameras capture visual signals, send video information, oversee operations, showcase data, and conduct computations. They find application in safety monitoring at workplaces, automated pipeline networks, self-driving vehicles, and smart transportation systems. In order to address the requirements of holistic understanding, widespread connectivity, and intelligent management of urban services, monitoring, tourist attractions, environmental issues, and long-term video surveillance, a significant deployment of smart sensors and Internet of Things (IoT) devices is necessary. However, these devices are frequently used in challenging environments and remote areas, which can restrict their capacity to offer consistent and dependable sustainable data communication and video event processing [52].

1.4.2 Smart Grid

Today's smart urban areas will necessitate a network capable of bolstering system dependability through the integration of innovative technologies. Moreover, it will facilitate comprehensive power management from start to finish, encompassing advanced utility planning operations. The current power grid is antiquated and requires enhanced systems to achieve superior, dependable, and cost-effective power distribution, data transmission, and secure integrated communication. The utilization of the smart grid, coupled with digital communications, enables continuous monitoring of the network's functionality. In the event of any identified network irregularities, the system can promptly rectify and restore them, ensuring seamless operation a process known as self-healing [53].

The power grid is encountering fresh obstacles amidst the rapid urbanization and the escalating need for more dependable and accessible power provision. As the call for cutting-edge technology to support efficient smart city infrastructure rises, there is an increasing desire for automation throughout the network. The notion of smart city power distribution, incorporating intelligent equipment solutions, has been intricately crafted to cater to the demands of urban power distribution settings. Ensuring dependable power distribution is essential for upholding regular daily activities and business functions, as well as guaranteeing that emergency services can access electricity [54, 55]. To secure uninterrupted power supply to critical zones like business hubs, it is vital to receive early alerts prior to any malfunctions and to possess a network that facilitates the organization of backup supply pathways. In times of urgency, it is vital to promptly and accurately pinpoint issues, segregate them, and reinstate functionality by remotely reconfiguring the network. Moreover, an increasing need exists for energy conservation, flexibility in meeting varying energy needs, incorporation of dispersed power generation, and facilitation of electric vehicle (EV) charging. These criteria highlight the importance of having instant access to up-to-date grid data. The idea of smart city (shown in Fig. 1.8) power distribution caters to the total energy needs of cities, offering a gradual and flexible strategy to attain the most feasible resolution [57].

1.4.2.1 Reliability

This technology utilizes state estimation to enhance error detection and enable self-recovery of the system without the need for technician intervention. This will ensure a more reliable supply of electrical energy, reducing vulnerability to natural disasters or attacks. While multiple routes are often praised as a feature of the smart grid system, the older technology also had several routes. The power lines in the grid were initially constructed using a branched model. Over time, the connectivity was improved by implementing multiple routes through a network structure. However, a new problem has arisen: if the current flow in the arrangement or related effects surpass the limits of the specific system constituent, it may not be able to handle it, and the current will be redirected to other network essentials, potentially leading to a domino effect of failures [58].

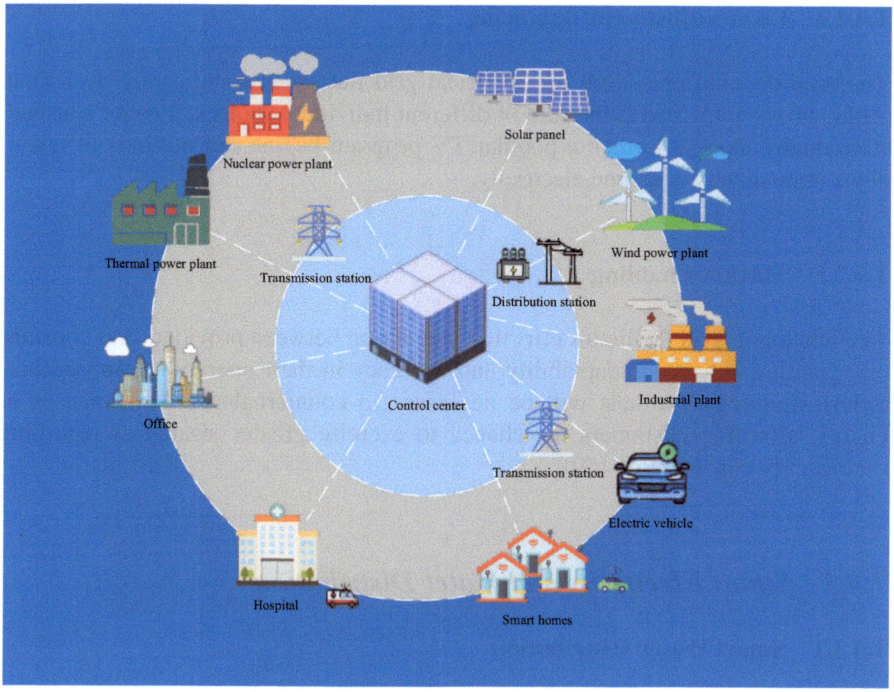

Fig. 1.8 Smart city and smart grid [56]

1.4.2.2 Flexibility in System Topology

In the future, the transmission infrastructure will be upgraded to accommodate bidirectional energy currents. This will enable the use of distributed generation technologies such as PV panels on rooftops, fuel cells (FCs), electric vehicle batteries, wind turbines, and other sources. Standard grids were originally designed to handle the direct flow of electricity. However, if a limited sub-network generates more energy than it consumes, the counterflow can potentially create reliability and safety concerns. A smart grid system is designed to address these situations [57].

1.4.2.3 Effectiveness

Improvements in power infrastructure efficiency can be achieved through the implementation of a smart grid system. This system includes consumer management strategies such as adjusting air conditioner usage during periods of high energy prices, optimizing voltage levels, eliminating the need for manual meter reading, and improving outage management through the use of data from advanced metering infrastructure systems. Typically, the impact is reduced redundancy in conduction and supply lines, resulting in improved generator utilization and lower power costs.

1.4.2.4 Load Adjustment/Balancing

The total demand placed on the electrical grid fluctuates significantly over time. Although the total load is the sum of different individual choices, it is not constant. It gradually increases when a popular TV program begins and millions of televisions immediately consume electricity.

1.4.2.5 Market Enabling

The intelligent grid facilitates effective interaction between providers and consumers, granting enhanced adaptability and accuracy in their operational approaches. Solely significant burdens will be necessary to counterbalance the expenses of energy, affording customers the chance to exercise greater awareness regarding their energy consumption [59].

1.4.3 Smart Electricity and Water Distribution

1.4.3.1 Smart Water Distribution

1.4.3.1.1 Definition of Smart Water Management (SWM)

Smart water management (SWM) is an essential component of sustainable urban water systems, with the goal of efficiently utilizing water at the regional or municipal level in a way that is environmentally friendly and independent. SWM employs cutting-edge technologies, including information technology, control, and monitoring systems, to minimize leakage, guarantee water quality, enhance customer satisfaction, streamline operations, and attain other significant performance advantages. A smart city fosters socioeconomic advancement and technological innovation by allocating resources to enhance human and social capital through the utilization of information and communication technology (ICT).

Smart water management (SWM) offers numerous advantages, such as enhanced comprehension of the water infrastructure, identification of leaks, promotion of water conservation, and monitoring of water quality. Additionally, it allows public utility companies to establish a comprehensive database capable of identifying locations where water loss or unauthorized connections take place. The advantages of smart water networks encompass financial savings in water and energy preservation, enhanced customer service, and potential reductions of up to 30% on water expenses [42].

The development of ICT is growing and becoming more advanced, enabling more efficient management and optimization of infrastructure for drinking water supply systems (SPAM) and centralized wastewater treatment plants (WWTP). Smart water refers to the advanced system of infrastructure that is designed to efficiently manage the utilization of water and energy resources required for their transportation. It effectively oversees the complete water cycle through the use of recycling technology, water treatment technology, and monitoring and control technology.

A cohesive and enduring drinking water supply system is a crucial element of intelligent water management, tailored to meet the various service requirements and consolidated into a single system to uphold the standards of water quality and quantity. Smart water technology allows drinking water providers to minimize nonrevenue water (NRW), which refers to the discrepancy between the amount of water produced and the amount that is actually sold or distributed. Figure 1.9 shows an integrated urban water management system [61].

1.4.3.1.2 Smart Water Grid

The smart water grid (SWG) is a novel water management system that leverages ICT to achieve maximum efficiency and sustainability in water resource control. The primary goal of the SWG is to efficiently distribute and oversee water resources by utilizing technologies like automated meter reading (AMR), advanced metering infrastructure (AMI), and supervisory control and data acquisition (SCADA). Sensus introduces an advanced method for monitoring water distribution networks using a bidirectional real. The SWG comprises five layers: measuring and sensing devices, wireless communication channels, database management software, real-time data analysis and modeling software, and automation and control tools. The objective is to activate the utility from a remote location and automatically monitor and control devices in the network, using SCADA systems to optimize control of the water distribution network [60]. Figure 1.10 displays a smart water grid.

1.4.3.1.3 Flood Early Warning System (FEWS)

Flood impact poses significant challenges in urban areas, where effective flood management strategies play a crucial role in mitigating economic losses and human casualties. Utilizing flood early warning systems (FEWS) emerges as a key solution in addressing flood impacts comprehensively, considering various societal and financial aspects. Analyzing urban FEWS extensively is essential for enhancing national economic development through improved flood preparedness and risk reduction efforts.

FEWS is founded upon an examination of inherent and geographical principles, taking into account human, societal, and economic attributes. The system employs suitable techniques for monitoring and simulating, forecasting information attributes associated with floods, and disseminating information to individuals. The primary objective of FEWS is to guarantee the security and protection of human life and assets.

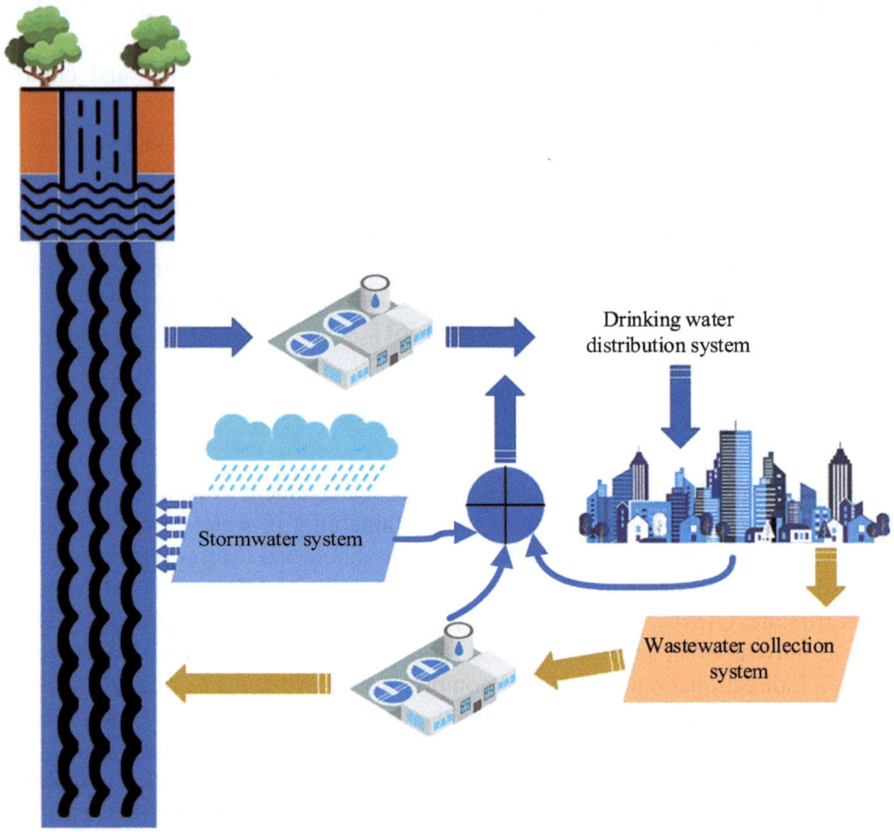

Fig. 1.9 Integrated urban water management system [60]

Scientists have suggested utilizing SWM, MIKE, and other models to replicate the depth and extent of flooding, as well as other indicators of flooding, in various scenarios. Integrating rainfall and inundation indexes for early warning can mitigate the influence of underlying surface and drainage systems on the accuracy of early warning systems.

Flood loss assessment can serve as a scientific foundation for enhancing urban flood warnings. However, relying solely on rainfall and inundation data fails to adequately capture the spatial variations in property damage within urban areas. This paper suggests the development of a "three information" early warning system that integrates data on rainfall, inundation, and losses to provide a comprehensive multi-information early warning. The primary objective of this system is to offer a scientific foundation and direction for decision-making related to disaster prevention and mitigation in flood-prone regions across the globe [63].

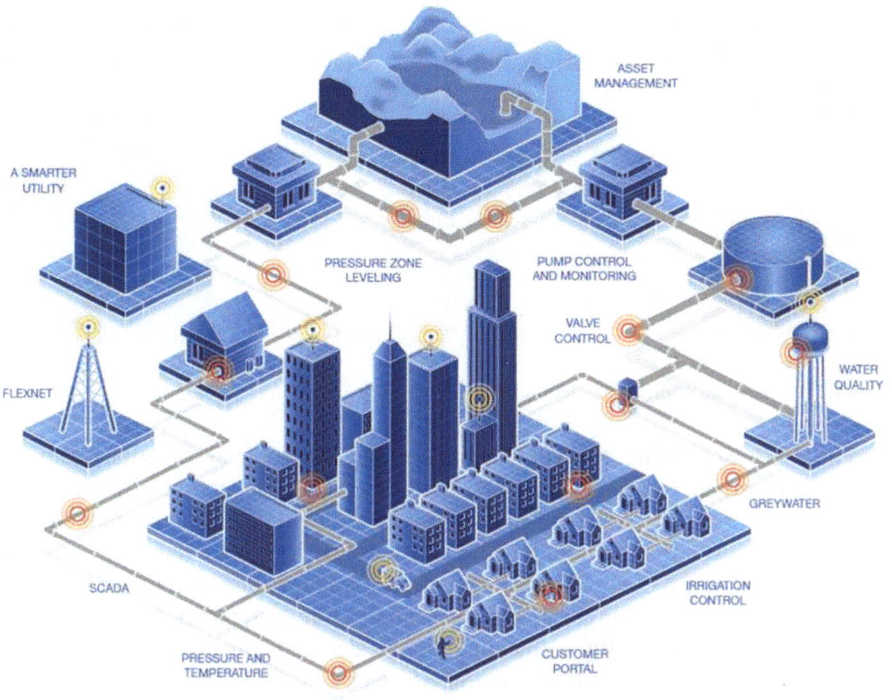

Fig. 1.10 Smart water grid diagram (figure is adopted from [62])

1.4.3.1.4 Water Quality Online Monitoring System (Onlimo)

The current situation calls for the implementation of measures that are aimed at reducing pollution and improving the quality of water bodies. This is so because the quality of the water in rivers and lakes is continuing to deteriorate. Constant and real-time deployment of online technology for the purpose of monitoring water quality is one approach that can be taken to address this problem. When it comes to the management of watersheds and the formulation of policies for the management of river and lake water on both a national and regional scale, the data that is obtained through the monitoring system can serve as an essential instrument for the development of solutions. Additionally, this information can be utilized to educate the community on the significance of conservation and protection efforts, as well as to raise public awareness about issues pertaining to water quality [64].

1.4.3.2 Smart Electricity Distribution

The notion of intelligent urban power distribution which is shown in Fig. 1.11, incorporating advanced equipment solutions, is tailored to address the needs of urban power distribution settings. Ensuring dependable power distribution is

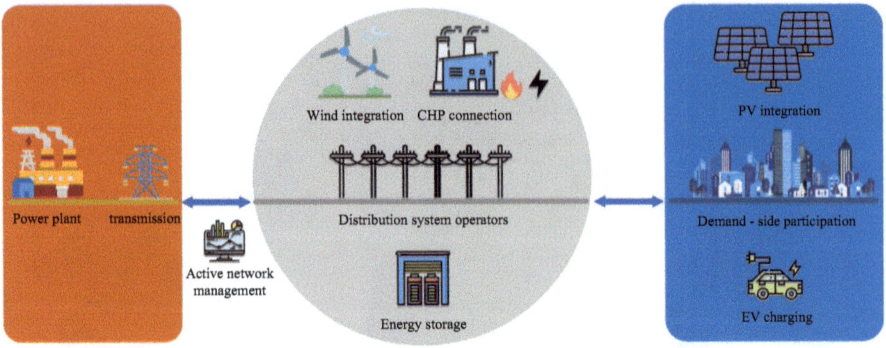

Fig. 1.11 Smart electricity distribution [65]

essential for sustaining everyday activities and commercial functions and for enabling emergency services to access electricity. Early detection of faults and a flexible network for establishing backup supply pathways are vital for maintaining continuous power distribution to critical zones like business hubs. In times of urgency, swift and accurate fault identification, isolation, and restoration of power via remote network reconfiguration are essential. Moreover, the increasing need for energy efficiency, flexibility in meeting energy demands, incorporation of decentralized power sources, and facilitation of electric vehicle charging are driving forces. These demands highlight the importance of real-time grid data access. The smart city power distribution concept caters to urban energy demands comprehensively, offering a gradual and flexible strategy to attain optimal solutions [66].

1.4.4 Smart Building

A smart building represents an advanced edifice that leverages cutting-edge technologies to interconnect, assess, and enhance its functionality. By harnessing state-of-the-art tools, smart buildings furnish proprietors with invaluable information to elevate operational efficiency, engineered for adept monitoring, regulation, and improvement of diverse operational facets, smart buildings commonly IoT and artificial intelligence (AI). Figure 1.12 shows elements of smart building technology [68]. Prominent components of smart buildings encompass:

1.4.4.1 Energy Efficiency

Smart building design reduces energy usage by implementing strategies like efficient heating, ventilation, and air conditioning (HVAC) systems, optimized lighting, and intelligent power management. It utilizes sensors and data analysis to optimize energy consumption according to occupancy and external conditions.

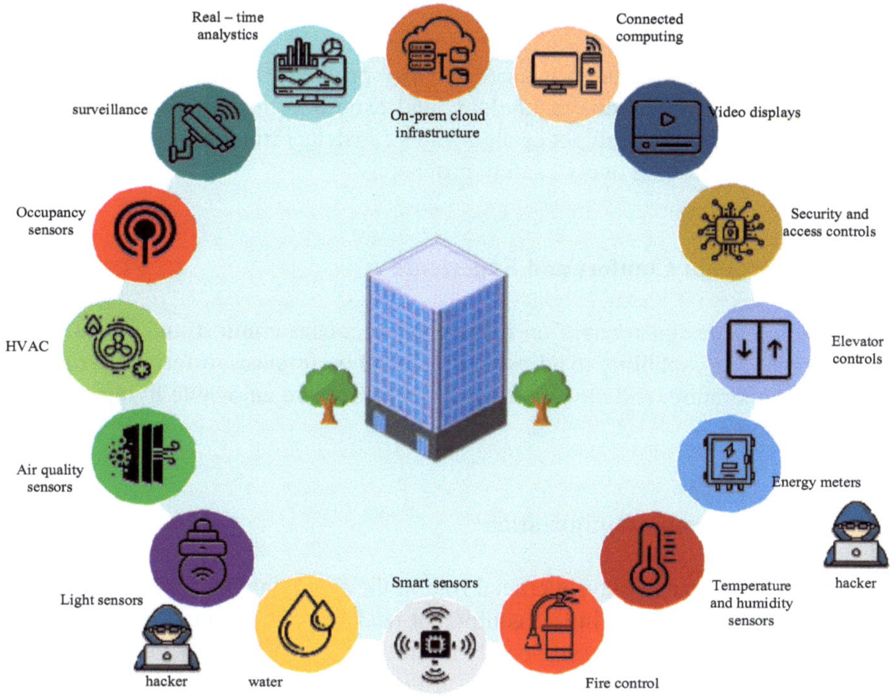

Fig. 1.12 Smart building technology [67]

1.4.4.2 Automated Systems

Smart building technology effortlessly combines various systems such as heating, ventilation, air conditioning, lighting, security, and access control within a centralized automation platform. This integration allows for the seamless and coordinated control of these systems, leading to improved effectiveness and reduced inefficiencies.

1.4.4.3 Data Analytics

Buildings with advanced technology collect data from diverse sensors, meters, and devices. This information undergoes thorough analysis to offer valuable perspectives on building efficiency, occupant actions, and potential enhancements to facilitate better decision-making.

1.4.4.4 Connectivity and Interoperability

Smart building technology involves the use of interconnected IoT devices and sensors that facilitate smooth communication and data exchange between different building systems and subsystems. This ensures that information is effectively communicated to enhance the overall efficiency of building operations.

1.4.4.5 Safety and Security

Intelligent structures feature advanced security protocols such as biometric entry control, modern surveillance technology, and extremely responsive fire detection mechanisms. Artificial intelligence algorithms can detect irregularities and possible risks, thereby improving overall safety provisions.

1.4.4.6 Occupant Comfort and Experience

Smart building design focuses on optimizing occupant comfort and productivity. Systems have the capability to adjust to individual preferences, offering customized lighting, temperature, and other settings to create a more enjoyable living or working environment.

1.4.4.7 Maintenance Optimization

Predictive maintenance plays a crucial role in the functionality of smart buildings. Through real-time monitoring of equipment health and performance, early detection of potential issues can significantly minimize downtime and repair expenses.

1.4.4.8 Sustainability

Smart constructions aid in creating a more eco-friendly tomorrow through reducing energy consumption, enhancing resource distribution, and integrating sustainable energy origins. This assists in lowering the carbon footprint and general environmental influence of constructions.

1.4.5 Smart Transportation

Smart transportation systems utilizing modern ICT and IoT infrastructures are increasingly employed to boost transportation infrastructure's safety and efficiency (Fig. 1.13). These systems tackle challenges like traffic congestion, noise pollution, and road accidents, resulting in reduced travel time and fuel consumption. Incorporating ICT alongside advanced sensors, computers, electronics, and communication technologies is crucial for creating an intelligent and sustainable transportation system. By integrating state-of-the-art features in communication, electronics, and computers, it becomes feasible to transmit information, regulate traffic flow, and manage transportation networks effectively. The adoption of new

Fig. 1.13 Smart transportation [69]

technologies in transportation systems prioritizes sustainability, integration, safety, and responsiveness as the core principles. The fundamental principles play a crucial role in achieving the main goals of intelligent transportation, such as enhancing accessibility and mobility, safeguarding the environment, and promoting economic progress. By integrating traffic control systems, operating efficiently on a large scale, and analyzing extensive data sets, smart transportation solutions can enhance safety for travelers and tackle urban population challenges. Advancements in technology will enable sustainable transportation infrastructures, promoting efficient utilization of current facilities for managing vehicle flow, leading to better congestion control and minimizing congestion-related drawbacks [70].

1.4.6 Smart Services

A smart city is a location where conventional networks and services are enhanced in efficiency by utilizing digital and communications technology, resulting in advantages for residents and businesses. The ISO family of standards prioritizes the utilization of data information and contemporary technology to provide enhanced services and improve the quality of life for inhabitants, businesses, and tourists. The involvement of end-users, including people, corporations, and other stakeholders, is essential in this setting. ICT is widely regarded as the primary facilitator of smart cities. However, the implementation and use of this technology in the form of smart services hold the greatest potential for sustainable growth and development. Effective leadership and managerial strategies play a vital role in the intricate setting of a smart city to guarantee the beneficial utilization and fair distribution of ICT advantages. The integration of smart services is imperative in this cohesive collaboration, establishing an environment where people, technology,

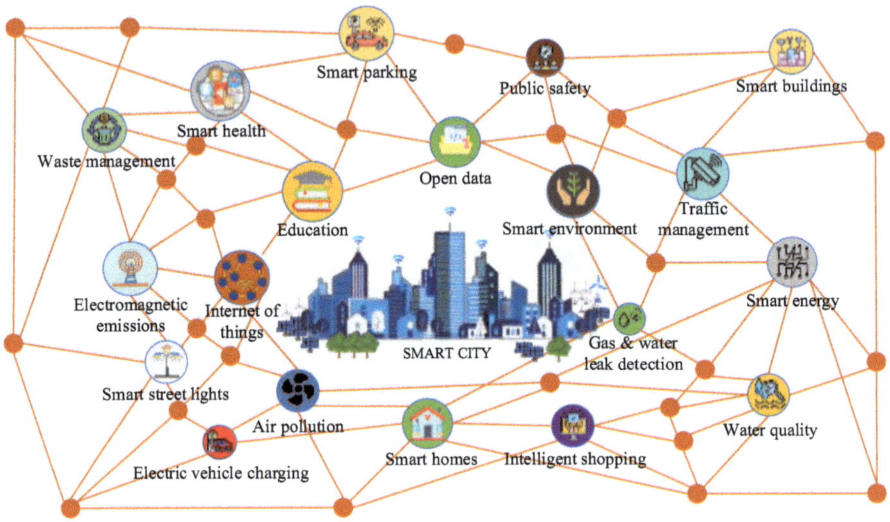

Fig. 1.14 Smart services [72]

entities, and data are seamlessly amalgamated through connectivity, communication, data acquisition, and engagement. Smart services are proactive and rely on up-to-date field data, enabling the delivery of services independent of any mishaps or failures. Although there is a significant amount of study on smart services, the literature has not adequately explored the direct relationship between smart service delivery and smart cities, particularly in terms of factors like as growth, development, sustainability, and citizens' well-being [71]. Figure 1.14 shows elements of smart services.

1.4.7 Smart Infrastructure

Infrastructure can be understood differently depending on the specific situation in which the term is used. In terms of usefulness and operational effectiveness, infrastructure includes underground and overhead networks of cables and pipelines, as well as all related resources. On the other hand, civil engineers concentrate on extra service features in urban regions, such as roadways, bridges, train and bus terminals, educational establishments like schools and colleges, medical centers such as hospitals, and a range of other communal facilities [73]. Figure 1.15 provides the classification of smart infrastructure development and monitoring.

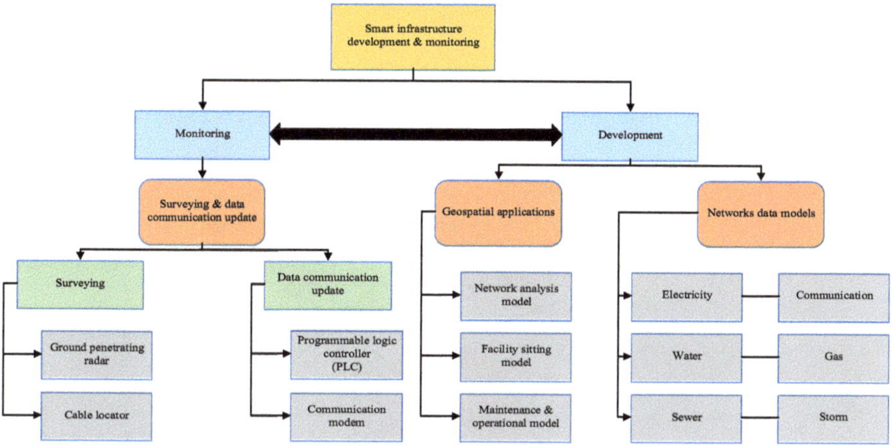

Fig. 1.15 Smart city infrastructures development and monitoring framework [73]

1.4.8 Information and Communication Technology (ICT)

ICT and communication technology are crucial for the progression of intelligent urban development and the promotion of long-term sustainability. ICT provides various tools that facilitate the collection, analysis, and dissemination of information regarding urban life, thus enhancing living standards and promoting sustainability. The efficient utilization of ICT has the potential to boost creativity and cultural expression in cities, while also fostering economic advancement and societal interconnectedness. Smart and sustainable cities integrate sustainable development, industrialization, and technological advancement, blending urban growth with a habitable environment.

Urban sustainability is an essential strategy for addressing urban development challenges and creating eco-friendly and inclusive cities in an increasingly urbanized world. This involves leveraging technology, creativity, and sustainable practices to enhance the quality of life for city dwellers, all the while mitigating adverse environmental impacts. ICT holds promise in optimizing urban transit systems, analyzing public safety data, fostering sustainability and resilience, boosting energy efficiency, and strengthening disaster readiness.

ICT is pivotal in the realm of transportation as it provides up-to-the-minute traffic information, supports sophisticated traffic control systems, reduces traffic jams, improves air quality, and fosters urban development. The development of intelligent cities is essential for enhancing public safety by deploying live monitoring in high-crime zones, intelligent security systems, and emergency services. Smart grid innovations boost energy efficiency, monitor energy usage patterns, and endorse sustainable energy resources, thus reducing carbon emissions and conserving natural reserves. Figure 1.16 illustrates the role of ICT in a sustainable smart city [75].

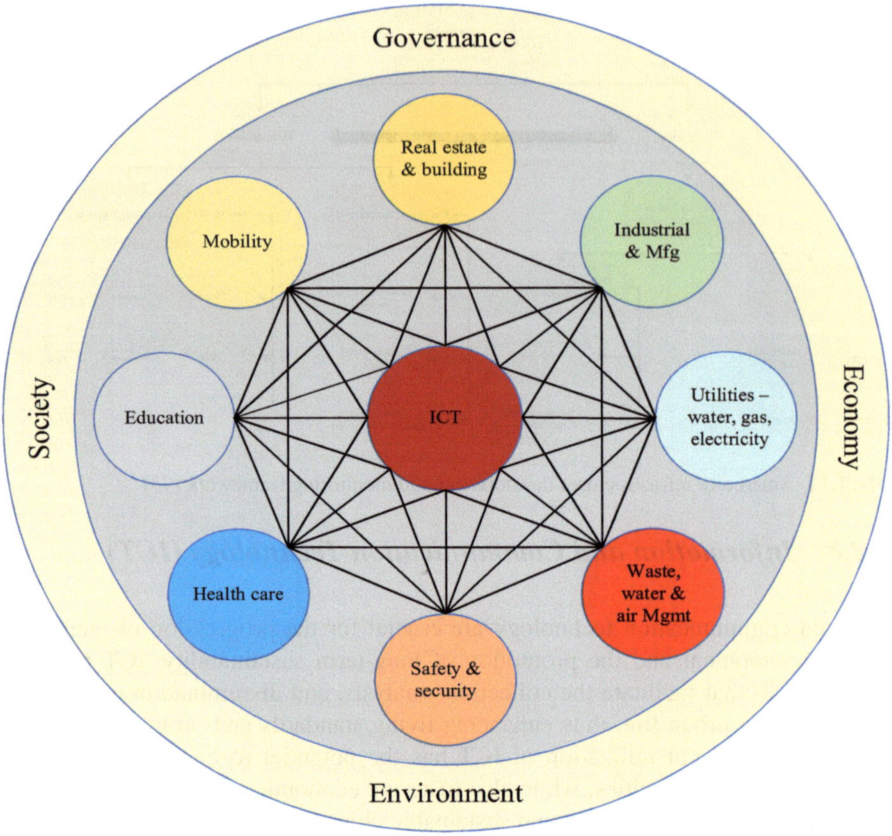

Fig. 1.16 Sustainable smart cities and position of ICTs [74]

1.5 Energy Storage

In an alternating current (AC) system, electrical energy cannot be stored electrically. Nevertheless, there exist multiple techniques for storing it by transforming AC energy into electromagnetic energy storage systems like superconducting magnetic energy storage (SMES), electrochemical methods such as various types of batteries (accumulators), kinetic storage (flywheels), or potential energy storage (hydropower plants) or compressed air [compressed air energy storage (CAES)]. The energy storage solutions currently accessible in the market include battery energy storage systems (BESS), energy capacitor systems (ECS), and flywheel energy storage systems (FESS). It is not feasible for energy storage systems (ESS) in an AC grid to directly store electrical energy.

The BESS is frequently utilized, yet it comes with limitations like restricted lifespan, current, and voltage constraints, as well as environmental risks. Due to the substantial advancement in renewable energy sources (RES), the progression of electromobility, and the necessity to enhance the operation of current power networks, the significance of electricity storage has surged in recent times [76].

1.5.1 Types of Energy Storages in Smart Cities

1.5.1.1 Batteries

1.5.1.1.1 Lead Acid

Lead-acid batteries, the most ancient form of electricity storage, are extensively utilized across multiple industries due to its uncomplicated design, ability to produce large levels of electric current, resistance to excessive charge, and affordability. Recent advancements have led to the creation of improved methods to enhance their functionality, such as maintenance-free batteries equipped with liquid electrolyte, controlled valves, glass mat separators, and gel electrolyte. Recent research indicates that substituting lead with lighter elements such as carbon could potentially enhance power and energy density.

1.5.1.1.2 Li-ion

Lithium-ion batteries, a specific type of battery, have become significant in the field of energy storage because of their elevated voltage and energy density. These batteries are utilized in household and system storage tanks, providing benefits like as high energy-to-weight ratios, lack of memory effect, and minimal self-discharge. These batteries are mostly employed in portable devices and hybrid and rechargeable electric cars; however, the initial expenses pose a substantial obstacle to their widespread adoption.

1.5.1.1.3 NiCd/NiMH

The development of energy storage technology in nickel-cadmium batteries has been a notable progress since the early 20th century, providing a quick charging time and durability against changes in temperature. These batteries were utilized in demanding applications such as radios, camera spotlights, and power tools. Nevertheless, they possess drawbacks such as the memory effect, elevated expense, and an intricate disposal procedure caused by poisonous substances. Nickel metal hydride batteries (NiMH) have become more popular than NiCd batteries because they have a greater power density, are safer, and have a longer lifespan when partially charged. NiMH batteries are extensively utilized in electric vehicles (EVs) and hybrid electric vehicles (HEVs) because of its superior power density, safety, and excellent ability to withstand mistreatment. Nevertheless, they exhibit a significant self-discharge rate, which may result in cell rupture or polarization. Although NiMH batteries have several disadvantages, they do have environmental advantages, including efficient recycling and the lack of harmful substances in the cells.

1.5.1.1.4 NaS

The process of energy storage in sodium-sulfur solid beta alumina utilizes a molten sulfur cathode and a molten sodium anode, separated by a nonporous solid beta alumina ceramic electrolyte. This storage solution, characterized by its high density, efficiency, and extended lifespan, enhances the performance of power grids and renewable energy power plants. It also allows for quick and seamless integration into operation.

1.5.1.1.5 FBs

Flow batteries, often known as FBs, are a highly promising technology that enables the separation of total energy stored from nominal power. The reactor and auxiliary tank's dimensions dictate the nominal power and stored capacity of the battery, rendering them appropriate for the substantial power and energy demands of electric utilities. An example of a widely used technology on Facebook is iron-chromium flow batteries (ICBs), which were created through a collaboration between NASA and Mitsui. ICBs possess superior efficiency, effortless scalability, and dependability, rendering them appropriate for energy storage systems with multimegawatt capacity as well as smaller uninterruptible power supply (UPS) systems. Despite the challenges posed by the low voltage and parasitic chemical reactions in these devices, their efficiency is significantly affected. Despite these hurdles, FB technology is currently utilized in a variety of demonstration and commercial products, ranging from large-scale to small-scale applications.

1.5.1.1.6 EDLCs

Supercapacitors provide exceptional efficiency and endurance owing to their elevated power density, which leads to rapid charge and discharge durations. They possess the capability to store electric energy directly, enabling rapid charging and discharging with exceptional efficiency. Supercapacitors are utilized in active filters, distribution networks, and electric vehicles to enhance electricity quality and optimize energy balancing.

1.5.1.2 Fuel Cells

Regenerative fuel cells (FCs) are devices that integrate the capabilities of a fuel cell (FC) and electrolyze into a single unit. They generate electricity by utilizing hydrogen stored in a gaseous fuel, which can subsequently be employed for energy storage. This novel strategy guarantees the establishment of a "hydrogen economy" by merging energy storage and electricity generation, resulting in a reduction in the size of systems that necessitate both functionalities.

1.5.1.2.1 Alkaline Fuel Cells

Alkaline fuel cells, often referred to as Bacon cells, employ a liquid alkaline electrolyte such as potassium hydroxide (KOH) and exhibit resilience in challenging environments. These devices find application in uninterruptible power supplies (UPS), telecommunications systems, and electric buses. Nevertheless, their susceptibility to carbon monoxide is so low that they are not feasible for extended periods of time.

1.5.1.2.2 Phosphoric Acid Fuel Cells

Phosphoric acid fuel cells (PAFCs), which were created in the 1950s, are a commercially available technology that offers enhanced operational stability, efficiency, and cost reduction. The electrolyte used is gel orthophosphoric acid, which is contained within a porous layer made of Teflon and silicon carbide. The electrodes are made of porous graphite with a platinum admixture. PAFCs function at elevated temperatures, rendering them exceptionally resilient to carbon monoxide pollution and attaining remarkable process efficiency.

1.5.1.2.3 Direct Methanol Fuel Cells

NASA has created fuel cells that provide a novel method for storing electricity by utilizing methanol. These fuel cells exhibit a high energy density and cheap production costs and provide easy transport and storage. They are appropriate for low-power loads like as portable electronic gadgets or power banks. Recent advancements have modified them to be used as power sources for compact crane trucks in expansive warehouses, resulting in decreased charging time and eliminating the need for battery installation expenses. Additional fuel cell technologies encompass molten carbonate, proton exchange membrane, and solid oxide.

1.5.1.3 Hydrogen Energy Storage

Hydrogen, the most prevalent element on Earth, functions as an unlimited energy source that can be transformed into heat or electricity. Two research studies by the U.S. National Research Council and the American Physical Society have examined the technological options and expenses associated with hydrogen. However, the main focus remains on the overall energy balance of an economy reliant on hydrogen, which involves various stages like production, compression, liquefaction, conveyance, distribution, and retention [77].

1.5.2 Benefits of Energy Storages

Hydrogen, the most abundant element on Earth, acts as an unlimited energy source that can be transformed into heat or electricity. Two research studies by the U.S. National Research Council and the American Physical Society have assessed the options and expenses associated with hydrogen. However, the main focus remains on the overall energy balance of an economy based on hydrogen, which involves various stages like production, compression, cooling, conveyance, distribution, and retention [78].

1.5.2.1 Reducing Peak Demand

Energy storage systems provide substantial benefits in mitigating peak demand on the power grid, alleviating strain, and decreasing the necessity for costly infrastructure enhancements. These devices offer supplementary power during moments of high consumption, thereby alleviating strain on the grid.

1.5.2.2 Enhancing Grid Reliability

Energy storage devices enhance grid stability by offering contingency power during blackouts or grid outages. During instances of power source disruption, such as natural disasters or equipment breakdowns, energy storage systems can offer a dependable and instantaneous power supply, guaranteeing the continued functioning of essential services.

1.5.2.3 Integrating Renewable Energy Sources

Renewable sources like solar and wind, subject to weather variations, require storage mechanisms to maintain a consistent power output. Energy storage systems are employed to manage electricity supply and demand, facilitating the incorporation of renewable energy sources into the energy grid.

1.5.2.4 Mitigating Power Outages

Energy storage systems offer a dependable and autonomous power supply, improving the standard of living and assisting vital services in isolated areas with restricted power grid availability.

1.5.2.5 Diverse Applications

Energy storage solutions play a vital role in enabling efficient storage and utilization of power, thereby improving the effectiveness and dependability of a wide range of energy-demanding operations.

1.5.2.6 Grid Stabilization and Peak Shaving

Energy storage systems maintain grid stability by equilibrating electricity supply and demand. They store surplus energy during periods of low demand and discharge it during periods of high demand, alleviating strain and averting power outages. Consequently, they optimize the distribution of energy.

1.5.2.7 Industrial and Commercial Applications

Storage systems enhance energy efficiency and reliability through the reduction of peak demand charges and the optimization of energy utilization across various applications.

1.5.2.8 Time Shifting

Energy resource storage improves electricity usage by storing surplus energy during periods of low demand, resulting in cost reduction and improved resource utilization during periods of high demand, ultimately leading to decreased electricity expenses.

1.5.2.9 Electric Vehicle Charging Infrastructure

Energy storage plays a crucial role in overseeing the energy of electric vehicles by enabling the storage of surplus energy. This practice alleviates strain on the power grid and enhances the vehicles' rapid charging capabilities.

1.5.2.10 Microgrid Support

Incorporating essential components into energy storage systems is vital for efficiently overseeing energy production and ensuring the robustness of microgrids, facilitating the delivery of reliable and stable electricity to communities, educational institutions, or remote areas.

1.5.2.11 Backup Power and Uninterrupted Power Supply (UPS)

Energy storage is essential for maintaining continuous power supply to vital services such as telecommunications during grid failures, hence minimizing significant repercussions.

1.5.2.12 Renewable Energy Integration

The efficient incorporation of renewable energy sources into the power grid via energy storage enables the retention of excess energy throughout periods of both high and low generation, so guaranteeing a steady and dependable supply [79].

1.6 Combined Heat and Power (CHP)

A CHP plant typically includes an electrical generator along with equipment for efficiently capturing and utilizing the heat generated by the generator which is shown in Fig. 1.17. The generator can be a prime mover, such as a gas turbine or a reciprocating engine. Alternatively, it could involve a steam turbine that generates power using high-pressure steam produced in a boiler. In certain situations, a CHP scheme can involve a combination of different components such as prime movers, boilers, and steam turbines [81].

1.6.1 CHP Configuration

1. The packaged CHP units are carefully planned and supplied as complete systems that seamlessly integrate with a building's electrical and heating systems. A steam boiler equipped with a steam turbine.
2. Custom-built CHP system meticulously crafted to perfectly match the unique needs of the location.
3. Micro-CHP presents an innovative approach with the goal of substituting conventional boilers in residential and small business settings.
4. Renewables CHP specifically designed to make use of sustainable fuels and feedstocks.

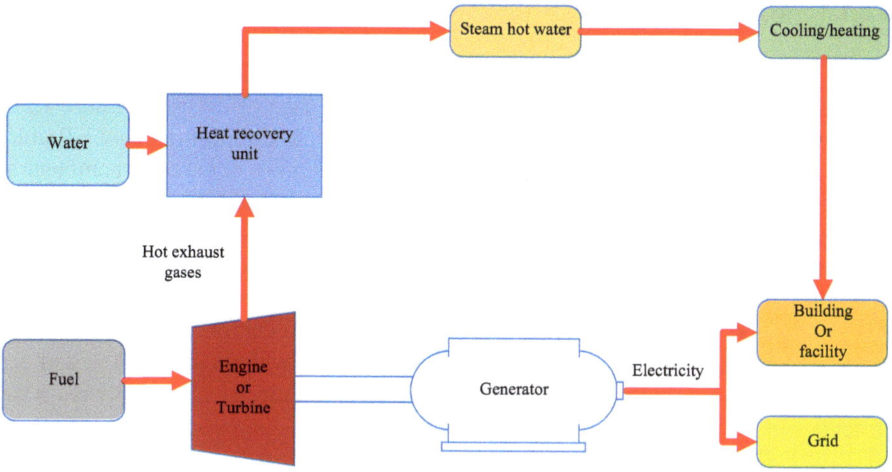

Fig. 1.17 Structure of CHP (water and heat) [80]

1.6.1.1 The Packaged CHP

Packaged CHP systems are designed and supplied as complete units that can be easily connected to a building's electrical and heating systems. Usually, these units come in various sizes, ranging from 50 kW to over 1 MW in electrical capacity. Typically, an integrated remote monitoring and control system is included.

One of the key benefits of the packaged CHP system is that it can be manufactured and prepared at the supplier's premises, making it convenient for delivery and positioning at the site. The majority of CHP packages are carefully crafted and provided as fully assembled units, tailored to fulfill the site's specific needs for reliability and availability in meeting its energy demands.

1.6.1.2 The Custom-Built CHP System

The custom CHP systems are meticulously crafted to perfectly align with the unique needs of each site. These systems are typically incorporated into the site's utilities and services. Usually, these schemes vary in size, ranging from generating 1 MWe to hundreds of MWe in generating capacity.

Custom CHP systems often utilize established technologies, with current trends leaning towards open and combined cycle turbine systems. These systems are preferred for their strong reliability and cost-effectiveness, especially at high generating capacities.

1.6.1.3 Micro-CHP

Micro-CHP refers to the generation of electricity on a smaller scale, typically producing less than 50 kW of electricity (kWe). In the case of domestic micro-CHP systems, they usually generate between 1 and 5 kWe. The initial micro-CHP units that were on the market had a preference for Stirling engines or external combustion engines. These units often have heat-to-power ratios that are significantly higher and heat capacities that are too large for most household uses. Nevertheless, there are now micro-CHP units that are utilizing fuel cell and internal combustion engine technologies. These are readily accessible with significantly lower heat capacities and heat-to-power ratios, making them a suitable choice for many households. Micro-CHP units can be used as direct replacements for domestic boilers and are eligible for support under the FiT scheme, as long as the electrical capacity is 2 kWe or less.

1.6.1.4 Renewable CHP

A key benefit of combined heat and power (CHP) generation is its capacity to harness and make use of a substantial quantity of heat that might otherwise be lost in conventional power generation techniques. This not only aids in cutting down on total fuel usage but also plays a part in lowering the release of greenhouse and harmful gases into the atmosphere. The utilization of carbon-neutral, renewable fuels in renewable CHP generation notably diminishes the carbon footprint of power generation.

1.6.2 Applications of CHP

CHP has served as a reliable energy solution for more than one hundred years, offering a multitude of advantages such as economic benefits, environmental sustainability, and improved dependability and robustness. It is particularly well-suited for establishments with steady requirements for power and heat all year round, as it has the capability to produce both electrical power and thermal energy. Numerous facility types have demonstrated the effectiveness of CHP, highlighting areas with further opportunities for development. These encompass:

1. Commercial buildings: Office buildings, hotels, nursing homes, and retail establishments
2. Residential: Residential structures, cooperative housing, and carefully designed neighborhoods
3. Institutions: Higher education institutions, medical facilities, correctional facilities, military installations

4. Municipal: Energy systems for districts, facilities for treating wastewater, and schools for K–12 education
5. Industrial: Agriculture, pulp, chemicals, paper, ethanol and food processing

1.6.3 Benefits of Combined Heat and Power in Smart Cities

Utilizing a heat engine or power station, a combined heat and power (CHP) system, also known as cogeneration, efficiently produces both electricity and heat simultaneously. The potential advantages of CHP from an economic standpoint are [82]:

- Reduced energy costs: CHP technology's notable effectiveness can result in substantial energy conservation compared to conventional separate purchases and on-site utilization of power and thermal energy systems.
- Offset capital costs: CHP can be installed as an alternative to boilers or chillers in new construction projects, or when there is a need to replace major heating, ventilation, and air conditioning (HVAC) equipment.
- Protection of revenue streams/reliability: By utilizing onsite generation and enhancing reliability, CHP ensures that critical infrastructure stays operational even during a disaster or significant power outage.
- Hedge against volatile energy prices: CHP could provide a solution to fluctuating energy prices by enabling users to generate their own power when electricity costs are high.

1.7 Benefits of Smart Cities

Smart cities employ information and communication technologies to improve service quality, promote citizen welfare, ensure sustainability, and foster economic growth. With the rapid growth of urban populations, the use of smart city technologies becomes essential for effectively and efficiently managing these populations [83].

- Smarter decision-making based on data analysis: City officials now have the ability to retrieve extensive volumes of big data using carefully planned data analytics methodologies. This enables them to discern places with a high-risk factor, predict the increase of the population, and recognize patterns in the interests of citizens. By continuously monitoring specific parameters in real time, the quality of service can be improved, resulting in cost reduction and enhanced services for residents. The fusion of big data and IoT presents limitless opportunities for enhanced decision-making.
- Increased involvement and interaction between citizens and the government: Residents are increasingly favoring smart cities because of their strong and user-friendly digital services. The services provided encompass various tools for

cooperation, applications for mobile devices, portals for self-service, and user-friendly online accounts. Enhancing digital services in communities enhances the appeal of smart cities, fostering a connected and seamless experience for citizens. Providing easily obtainable government data, interactive maps, and openness in budgeting fosters stronger connections with citizens, hence enhancing civic participation and fostering trust in city authorities.

- Community safety: A smart city is inherently more secure. Utilizing technological advancements and actively seeking collaborations between business and public sectors might effectively decrease criminal behavior. Law enforcement benefits from several technologies that provide them with a competitive advantage throughout their duties, including license plate recognition, gunshot detectors, connected crime centers, next-generation 911, and body cameras. Several communities have already commenced investing in intelligent technologies to facilitate the promotion of a more secure community.
- Minimized environmental effects: Smart cities are combatting pollutants and greenhouse gases by utilizing energy-efficient buildings, air quality sensors, and sustainable energy resources. These innovations assist in reducing the environmental impact in crowded urban regions by enabling the execution of strategies and tracking air quality metrics. The sensors possess the capacity to identify pollution origins and provide analytical insights, leading to diminished air pollution levels and the safeguarding of biodiversity. This holds significant importance as health problems linked to pollution contribute to the mortality of numerous individuals annually.
- Improved transportation: Urban transportation is undergoing a significant transformation due to the integration of intelligent technologies. These innovations, including smart traffic signals and advanced parking systems, play a crucial role in enhancing operational efficiency and reducing traffic congestion within cities. By facilitating better traffic control and enabling real-time monitoring of public transportation routes, these advancements help alleviate congestion during peak travel times. Forecasts indicate a projected annual growth rate exceeding 25% in smart urban transportation investments over the upcoming 5 years, potentially resulting in expanded land utilization for infrastructure development.
- Enhanced digital fairness: Smart city technology has the capacity to establish a fairer environment for inhabitants. In order to guarantee digital equity, it is imperative that citizens have access to high-speed internet services and cheap gadgets. Deploying strategically positioned public Wi-Fi hotspots across a city can provide dependable internet access to all its citizens.
- Emerging prospects for economic growth: An up-to-date examination highlights the significant impact of public funding in intelligent urban technologies on the expansion of a city's gross domestic product. This funding, backed by prominent private sector companies and municipal authorities, enhances the region's and the globe's competitiveness, thereby drawing in fresh residents and businesses. Integrated intelligent urban technologies enable enterprises to make informed choices through the utilization of accessible data platforms.

- Effective public utilities: Urban areas are undergoing a transformation due to the implementation of intelligent technologies that reduce wastage and conserve resources. These advancements support the swift identification of pipe leaks and prompt repairs, leading to a decrease in water wastage. Additionally, smart electric grids establish two-way communication between power suppliers and users, enhancing the detection of peak consumption periods and power failures.
- Improved infrastructure: Smart technology enables cities to detect and resolve problems in their infrastructure, such as deteriorating roads, bridges, and buildings. Through the application of predictive analytics, intelligent sensors can transmit data indicating alterations in the structure such as tilts or cracks, enabling prompt inspections and maintenance. This ultimately leads to cost savings for taxpayers and prevents infrastructure failures that could result in loss of life.
- Enhanced employee involvement and commitment: A smart city is crucial for a city to thrive. An efficient workforce is crucial for successfully implementing smart technologies. With the progress made in autonomous agents, mobile devices, bots, and sensors, city employees can now dedicate their attention to strategic initiatives, making manual processes more efficient and delivering better services. Implementing smart technologies necessitates foresight, financial commitment, and meticulous strategizing. Embracing the vision of a smart city is a wise choice [84–86].

1.8 Conclusion

The integration of renewable energy sources and the adoption of sustainable practices are crucial for the development of smart cities. Utilizing technologies such as smart grids, smart buildings, and intelligent urban power distribution systems can significantly enhance energy efficiency, reduce environmental impact, and improve the overall quality of life for urban residents. Furthermore, the shift towards renewable energy sources not only addresses the imminent challenges posed by the depletion of finite resources but also mitigates the environmental consequences associated with the burning of fossil fuels. By harnessing natural processes such as solar radiation, wind movement, and bioenergy, cities can transition towards a cleaner and more sustainable energy paradigm. Also, the promotion of sustainable energy resources in smart cities is not only a necessity in the face of increasing energy demands and environmental concerns but also a pathway towards creating resilient, efficient, and livable urban environments for current and future generations. Embracing renewable energy sources and implementing smart technologies are essential steps towards building a more sustainable and prosperous future for urban communities worldwide.

References

1. V. Todeschi, G. Mutani, L. Baima, M. Nigra, M. Robiglio, Smart solutions for sustainable cities—The re-coding experience for harnessing the potential of urban rooftops. Appl. Sci. **10**(20), 7112 (2020)
2. T. Cortese, J. Almeida, G. Batista, J. Storopoli, A. Liu, T. Yigitcanlar, Understanding sustainable energy in the context of smart cities: A PRISMA review. Energies **15**, 2382 (2022)
3. S.G. Malla et al., Coordinated power management and control of renewable energy sources based smart grid. Int. J. Emerg. Electr. Power Syst. **23**(2), 261–276 (2022)
4. M. Kumar, Social, Economic, and Environmental Impacts of Renewable Energy Resources, in *Wind Solar Hybrid Renewable Energy System*, vol. 1, (IntechOpen, London, 2020)
5. M.R. Patel, O. Beik, *Wind and Solar Power Systems: Design, Analysis, and Operation* (CRC Press, Boca Raton, 2021)
6. P. Sadorsky, Wind energy for sustainable development: Driving factors and future outlook. J. Clean. Prod. **289**, 125779 (2021)
7. L.J. Müller et al., The carbon footprint of the carbon feedstock CO_2. Energy Environ. Sci. **13**(9), 2979–2992 (2020)
8. M. Soltani et al., Environmental, economic, and social impacts of geothermal energy systems. Renew. Sust. Energ. Rev. **140**, 110750 (2021)
9. A. Baba, D. Chandrasekharam, Geothermal resources for sustainable development: A case study. Int. J. Energy Res. **46**(14), 20501–20518 (2022)
10. R. Archer, Geothermal energy, in *Future energy*, (Elsevier, Amsterdam, 2020), pp. 431–445
11. Y. Huang et al., A review of strategies for mitigating roadside air pollution in urban street canyons. Environ. Pollut. **280**, 116971 (2021)
12. M.A. Walton, Energy Security and the Energy Transition, in *Handbook on the Water-Energy-Food Nexus*, (Edward Elgar Publishing, Cheltenham, 2022), pp. 81–95
13. M. Foss, The impact of renewables on energy security. Canadian Foreign Policy J. **28**(3), 234–248 (2022)
14. M. Salimi, S.G. Al-Ghamdi, Climate change impacts on critical urban infrastructure and urban resiliency strategies for the Middle East. Sustain. Cities Soc. **54**, 101948 (2020)
15. D. Mahmood, N. Javaid, G. Ahmed, S. Khan, V. Monteiro, A review on optimization strategies integrating renewable energy sources focusing uncertainty factor–Paving path to eco-friendly smart cities. Sustain. Comput. Inf. Syst. **30**, 100559 (2021)
16. D. Apfel, S. Haag, C. Herbes, Research agendas on renewable energies in the Global South: A systematic literature review. Renew. Sust. Energ. Rev. **148**, 111228 (2021)
17. J. Zhao et al., The determinants of renewable energy sources for the fueling of green and sustainable economy. Energy **238**, 122029 (2022)
18. M.M. Hayati, H. Majidi-Gharehnaz, H. Biabani, A. Aminlou, M. Abapour, An Overview of Energy and Exergy Analysis for Green Hydrogen Power Systems, in *Green Hydrogen in Power Systems*, ed. by V. Vahidinasab, B. Mohammadi-Ivatloo, J. Shiun Lim, (Springer International Publishing, Cham, 2024), pp. 1–24
19. B. Li, Effective energy utilization through economic development for sustainable management in smart cities. Energy Rep. **8**, 4975–4987 (2022)
20. M.A. Ahad, S. Paiva, G. Tripathi, N. Feroz, Enabling technologies and sustainable smart cities. Sustain. Cities Soc. **61**, 102301 (2020)
21. P.A. Østergaard, N. Duic, Y. Noorollahi, S. Kalogirou, *Latest Progress in Sustainable Development Using Renewable Energy Technology*, vol 162 (Elsevier, Amsterdam, 2020), pp. 1554–1562
22. P. Pandiyan, S. Saravanan, K. Usha, R. Kannadasan, M.H. Alsharif, M.-K. Kim, Technological advancements toward smart energy management in smart cities. Energy Rep. **10**, 648–677 (2023)

23. E. Ghorbani, T. Fluechter, L. Calvet, M. Ammouriova, J. Panadero, A.A. Juan, Optimizing energy consumption in smart cities' mobility: Electric vehicles, algorithms, and collaborative economy. Energies **16**(3), 1268 (2023)
24. L. Jebaraj, A. Khang, V. Chandrasekar, A.R. Pravin, K. Sriram, Smart City: Concepts, Models, Technologies and Applications, in *Smart Cities*, (CRC Press, Boca Raton, 2023), pp. 1–20
25. D. Tang, Y.A. Solangi, Fostering a sustainable energy future to combat climate change: EESG impacts of green economy transitions. Processes **11**(5), 1548 (2023)
26. F. Sharipov, R. Omonboyev, Energy efficiency is an important factor of sustainable energy supply. Theor. Aspects Form. Pedag. Sci. **2**(2), 211–215 (2023)
27. F.S. Hafez et al., Energy efficiency in sustainable buildings: a systematic review with taxonomy, challenges, motivations, methodological aspects, recommendations, and pathways for future research. Energ. Strat. Rev. **45**, 101013 (2023)
28. J. Chen, Y. Zhang, S. Teng, Y. Chen, H. Zhang, F.-Y. Wang, ACP-based energy-efficient schemes for sustainable intelligent transportation systems. IEEE Trans. Intell. Vehicles **8**(5), 3224–3227 (2023)
29. K. Parham, M. Farmanbar, C. Rong, O. Arild, Assessing the Importance of Energy Management in Smart Homes, in *Proceedings of the 4th International Conference on Viable Energy Trends (InVEnT-2019)*, (ACM Digital Library, New York, 2019), pp. 26–28
30. T.M. Chen, Smart grids, smart cities need better networks. IEEE Netw. **24**(2), 2–3 (2010)
31. M.J.B. Kabeyi, O.A. Olanrewaju, Sustainable energy transition for renewable and low carbon grid electricity generation and supply. Front. Energy Res. **9**, 743114 (2022)
32. K. Tromly, *Renewable Energy: An Overview* (U.S. Department of Energy, Washington, 2001)
33. A.T. Hoang, X.P. Nguyen, Integrating renewable sources into energy system for smart city as a sagacious strategy towards clean and sustainable process. J. Clean. Prod. **305**, 127161 (2021)
34. F.M. Guangul, G.T. Chala, Solar energy as renewable energy source: SWOT analysis, in *2019 4th MEC international conference on big data and smart city (ICBDSC)*, (IEEE, Piscataway, 2019), pp. 1–5
35. A.G. Aberle, Surface passivation of crystalline silicon solar cells: A review. Prog. Photovolt. Res. Appl. **8**(5), 473–487 (2000)
36. V.S. Arutyunov, G.V. Lisichkin, Energy resources of the 21st century: problems and forecasts. Can renewable energy sources replace fossil fuels. Russ. Chem. Rev. **86**(8), 777 (2017)
37. A. Aslani, P. Helo, B. Feng, E. Antila, E. Hiltunen, Renewable energy supply chain in Ostrobothnia region and Vaasa city: Innovative framework. Renew. Sust. Energ. Rev. **23**, 405–411 (2013)
38. M. Kiraga, Hydroelectric power plants and river morphodynamic processes. J. Ecol. Eng. **22**(7), 163–178 (2021)
39. M.F.B. Alam et al., Assessing the factors influencing the adoption of geothermal energy to support the national grid in emerging economies: Implications for sustainability. Sustain. Oper. Comput. **5**, 167–180 (2024)
40. A.S. Gutiérrez, J.J.C. Eras, L. Hens, C. Vandecasteele, The energy potential of agriculture, agroindustrial, livestock, and slaughterhouse biomass wastes through direct combustion and anaerobic digestion. The case of Colombia. J. Clean. Prod. **269**, 122317 (2020)
41. M.J. Stolarski, K. Warmiński, M. Krzyżaniak, E. Olba-Zięty, M. Akincza, Bioenergy technologies and biomass potential vary in Northern European countries. Renew. Sust. Energ. Rev. **133**, 110238 (2020)
42. M.M. Hayati, A. Safari, M. Nazari-Heris, A. Oshnoei, Hydrogen-Incorporated Sector-Coupled Smart Grids: A Systematic Review and Future Concepts, in *Green Hydrogen in Power Systems*, ed. by V. Vahidinasab, B. Mohammadi-Ivatloo, J. Shiun Lim, (Springer International Publishing, Cham, 2024), pp. 25–58
43. A. Safari, M.M. Hayati, M. Nazari-Heris, Hydrogen-Combined Smart Electrical Power Systems: An Overview of United States Projects, in *Green Hydrogen in Power Systems*, ed. by V. Vahidinasab, B. Mohammadi-Ivatloo, J. Shiun Lim, (Springer International Publishing, Cham, 2024), pp. 321–340

44. H. Majidi-Gharehnaz, H. Biabani, A. Aminlou, M.M. Hayati, M. Abapour, Integration of Solar PV and GH2 in the Future Power Systems, in *Green Hydrogen in Power Systems*, ed. by V. Vahidinasab, B. Mohammadi-Ivatloo, J. Shiun Lim, (Springer International Publishing, Cham, 2024), pp. 203–224
45. A. Pareek, R. Dom, J. Gupta, J. Chandran, V. Adepu, P.H. Borse, Insights into renewable hydrogen energy: Recent advances and prospects. Mater. Sci. Energy Technol. **3**, 319–327 (2020)
46. C. Mukherjee, J. Denney, E.G. Mbonimpa, J. Slagley, R. Bhowmik, A review on municipal solid waste-to-energy trends in the USA. Renew. Sust. Energ. Rev. **119**, 109512 (2020)
47. M. Akcin, A. Kaygusuz, A. Karabiber, S. Alagoz, B.B. Alagoz, C. Keles, Opportunities for Energy Efficiency in Smart Cities, in *2016 4th International Istanbul Smart Grid Congress And Fair (ICSG)*, (IEEE, Piscataway, 2016), pp. 1–5
48. M. Kashef, A. Visvizi, O. Troisi, Smart city as a smart service system: Human-computer interaction and smart city surveillance systems. Comput. Hum. Behav. **124**, 106923 (2021)
49. A. Zanella, N. Bui, A. Castellani, L. Vangelista, M. Zorzi, Internet of things for smart cities. IEEE Internet Things J. **1**(1), 22–32 (2014)
50. P. Delaney, C. Pettit, Urban data hubs supporting smart cities. Research **14**, 7–9 (2014)
51. R. Massobrio, S. Nesmachnow, A. Tchernykh, A. Avetisyan, G. Radchenko, Towards a cloud computing paradigm for big data analysis in smart cities. Program. Comput. Softw. **44**, 181–189 (2018)
52. Y. Jin, Z. Qian, W. Yang, UAV cluster-based video surveillance system optimization in heterogeneous communication of smart cities. IEEE Access **8**, 55654–55664 (2020)
53. H. Shahinzadeh, H. Sadrarhami, M.M. Hayati, H. Majidi-Gharehnaz, M. Abapour, G.B. Gharehpetian, Review and Comparative Analysis of Deep Learning Techniques for Smart Grid Load Forecasting, in *2024 20th CSI International Symposium on Artificial Intelligence and Signal Processing (AISP)*, (IEEE, Piscataway, 2024), pp. 1–9
54. M.M. Hayati, B. Motallebi Azar, A. Aminlou, M. Abapour, K. Zare, Techno-Economic Analysis for Centralized GH2 Power Systems, in *Green Hydrogen in Power Systems*, ed. by V. Vahidinasab, B. Mohammadi-Ivatloo, J. Shiun Lim, (Springer International Publishing, Cham, 2024), pp. 59–83
55. A. Aminlou, M.M. Hayati, H. Majidi-Garehnaz, H. Biabani, K. Zare, M. Abapour, Techno-Economic Analysis for Decentralized GH2 Power Systems, in *Green Hydrogen in Power Systems*, ed. by V. Vahidinasab, B. Mohammadi-Ivatloo, J. Shiun Lim, (Springer International Publishing, Cham, 2024), pp. 85–103
56. C. Calcada. How Smart Grid Technology is Shaping Our Future. https://www.smorescience.com/how-smart-grid-technology-is-shaping-our-future/ (Accessed June 10, 2024)
57. O.S. Neffati et al., Migrating from traditional grid to smart grid in smart cities promoted in developing country. Sustain Energy Technol Assess **45**, 101125 (2021)
58. C. Mahapatra, Z. Sheng, V.C. Leung, T. Stouraitis, A Reliable and Energy Efficient IoT Data Transmission Scheme for Smart Cities Based on Redundant Residue Based Error Correction Coding, in *2015 12th Annual IEEE International Conference on Sensing, Communication, and Networking-Workshops (SECON Workshops)*, (IEEE, Piscataway, 2015), pp. 1–6
59. A. Aminlou, M.M. Hayati, K. Zare, P2P Energy Trading in a Community of Individual Consumers with the Presence of Central Shared Battery Energy Storage, in *Demand-Side Peer-to-Peer Energy Trading*, ed. by V. Vahidinasab, B. Mohammadi-Ivatloo, (Springer International Publishing, Cham, 2023), pp. 143–159
60. T. Hernaningsih, N. Said, S. Yudo, H. Wahyono, W. Widayat, A. Rifai, Application of the concept of smart city and smart water management for the new capital city. IOP Conf. Ser. Earth Environ. Sci. **1201**(1), 012103 (2023)
61. Z. Yehia. Non-revenue Water Reduction. 2023, Qeios, CC-BY 4.0, https://doi.org/10.32388/UP5M54.3, (Accessed June 10, 2024)
62. H.M. Ramos, A. Carravetta, A.M. Nabola, *New Challenges in Water Systems*, vol 12 (MDPI, Basel, 2020), p. 2340

63. Y. Zang, Y. Meng, X. Guan, H. Lv, D. Yan, Study on urban flood early warning system considering flood loss. Int. J. Disaster Risk Reduct. **77**, 103042 (2022)
64. S. Yudo, H. Wahyono, K. Alhasa, S. Zulaikha, Y. Pandiangan, Water quality data analysis from the onlimo monitoring system at 16 stations in priority watersheds and lakes. IOP Conf. Ser. Earth Environ. Sci. **1201**(1), 012052 (2023)
65. P.Q. Blog. What is a DSO? https://powerquality.blog/2022/01/21/what-is-a-dso/ (Accessed June 11, 2024)
66. Smart City Power Distribution_758009_LRENb, Report, ABB, https://shorturl.at/o7TTn, (Accessed June 11, 2024)
67. The Importance of Protecting Smart Building Technology from Cyber Threats. https://www.veridify.com/the-importance-of-protecting-smart-buildingtechnology-from-cyber-threats, (Accessed June 11, 2020)
68. S.B. Mathur, Smart grid – for smart cities, presented at the National Conference on Road Map for Smart Cities of Rajasthan, April 2017
69. R. Haigh, in is the MBA still relevant in a transforming world? ed, 26 november 2018
70. D. Oladimeji, K. Gupta, N.A. Kose, K. Gundogan, L. Ge, F. Liang, Smart transportation: an overview of technologies and applications. Sensors **23**(8), 3880 (2023)
71. R. Malik, A. Visvizi, O. Troisi, M. Grimaldi, Smart services in smart cities: insights from science mapping analysis. Sustainability **14**(11), 6506 (2022)
72. Polish investors invited to smart city projects in Tehran", February 21, 2018, https://www.tehrantimes.com/news/421460/Polish-investors-invited-tosmart-city-projects-in-Tehran, (Accessed June 11, 2024)
73. M. Al-Hader, A. Rodzi, The smart city infrastructure development & monitoring. Theor. Emp. Res. Urban Manage. **4**(2), 87–94 (2009)
74. A. Haidine, S. El Hassani, A. Aqqal, A. El Hannani, The role of communication technologies in building future smart cities. Smart Cities Technol. **1**, 1–24 (2016)
75. A.A. Batabyal, H. Beladi, The optimal provision of information and communication technologies in smart cities. Technol. Forecast. Soc. Chang. **147**, 216–220 (2019)
76. A. Colmenar-Santos, E.-L. Molina-Ibáñez, E. Rosales-Asensio, Á. López-Rey, Technical approach for the inclusion of superconducting magnetic energy storage in a smart city. Energy **158**, 1080–1091 (2018)
77. D. Rasolomampionona, M. Kłos, Energy Storage Systems and Their Role in Smart Grids, in *Smart Grids Technology and Applications*, (IntechOpen, London, 2023)
78. A.Z.A. Shaqsi, K. Sopian, A. Al-Hinai, Review of energy storage services, applications, limitations, and benefits. Energy Rep. **6**, 288–306 (2020)
79. Benefits and Applications of Energy Storage Systems. https://www.veolia.co.uk/benefits-and-applications-energy-storage-systems. (Accessed June 11, 2024)
80. U. S. E. P. Agency, in Combined Heat and Power (CHP) Partnership vol. 04/17/2024, ed
81. A detailed guide for CHP developers – Part 2. Department for Business, Energy & Industrial Strategy, February 2021
82. What are the benefits of a combined heat and power (CHP) system vol. 4/17/2024, ed: performance Services
83. M.W.S. Mansker-Young, Thinking about becoming a smart city? 10 benefits of smart cities. vol. 04/19/2024, ed: plante moran, July 14, 2023
84. A. Wirsbinna, L. Grega, Assessment of economic benefits of smart city initiatives. Cuad. Econ. **44**(126), 45–56 (2021)
85. R.P. Dameri, R. Garelli, Measuring Business Benefits and Performance in Smart Cities, in *European Conference on Innovation and Entrepreneurship*, (Academic Conferences International Limited, Reading, 2014), p. 137
86. G. Halegoua, *Smart Cities*; MIT Press: Cambridge, MA, USA, 2020.

Chapter 2
A Review of Geothermal Energy Resources and Applications in Jordan: Current Status and Prospects

Heba Alzaben

2.1 Introduction

Jordan relies heavily on imported oil and gas for the majority of its energy sources which pose a significant economic challenge in the country [1]. The amount of imported fossil fuel in Jordan has reached 94% [2]. Despite the constraints associated with Jordan's conventional energy sources, the rising demand for energy persists due to population growth and industrial expansion [3]. It is expected that Jordan's total energy demand will reach to 41.8 and 82.4 TWh in 2030 and 2050, respectively [2]. The use of fossil fuel energy resources is emerging as a notable environmental issue in Jordan. Therefore, it is important to seek sustainable and renewable energy solutions to address various environmental impacts associated with the use of fossil fuel sources including air, water, and plastic pollution [4]. Different research studies worldwide have extensively investigated the potential for utilizing renewable and sustainable energy sources including solar, wind, hydropower, biomass, and geothermal energy [5]. Renewable sources are considered as clean, unlimited, rapidly replenished naturally, and environmentally friendly resources that produce electrical power with much less greenhouse gas (GHG) emissions compared to fossil fuel sources [6]. Jordan's total renewable energy production in GWh over different years is summarized in Fig. 2.1 (the data in the figure are adapted from the International Renewable Energy Agency [IRENA] 2023 report).

One of the most important renewable energy sources is geothermal energy. It is an inexhaustible clean, and environmentally friendly, available 24 hours a day,

H. Alzaben (✉)
Faculty of Engineering, Mechanical Engineering, Al Hussein Technical University, Amman, Jordan
e-mail: heba.alzaben@htu.edu.jo

A. Ahmadian et al. (eds.), *Sustainable Energy Resources in Smart Cities*, Green Energy and Technology, https://doi.org/10.1007/978-3-031-91516-1_2

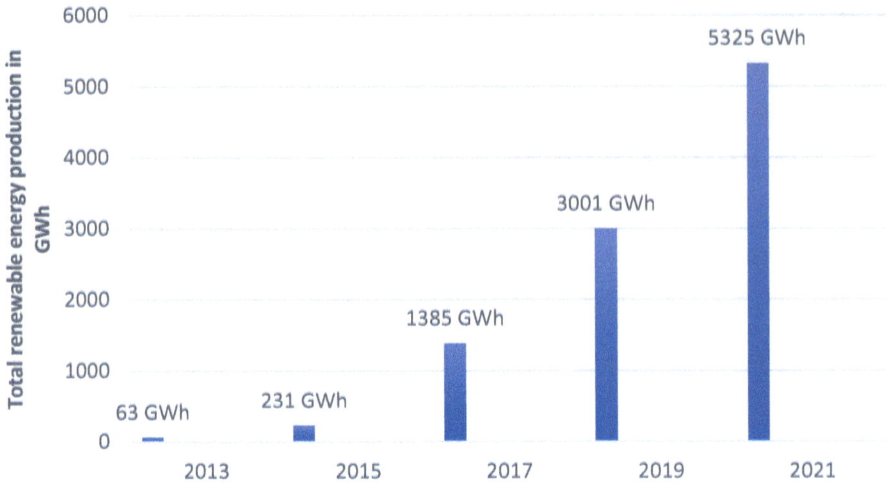

Fig. 2.1 Total renewable energy production in Jordan in GWh among different years (adapted from the International Renewable Energy Agency (IRENA) 2023 report)

365 days per year [7]. Geothermal word is composed of two terms "geo" which refers to the Earth and "thermal" which refers to heat and more specifically the underground heat [8, 9]. Geothermal energy is defined as the thermal energy generated and stored in the Earth [10]. The two primary sources of geothermal energy are the heat flow from the Earth's core and the thermal energy produced by the gradual decay of radioactive isotopes within the Earth's crust [11]. The geothermal gradient represent the rate of temperature increase with depth in the Earth's crust or in deep reservoirs, primarily governed by convective heat transfer [12]. Geothermal energy holds significant potential for medium temperature heat transfer applications [8], and it offer several advantageous characteristics, including its ability to provide baseload power, resilience to seasonal and weather related fluctuations, and stability against the impacts of climate change [13], suitablity for integration into both centralized and decentralized power generation systems and the widespread availability of resources across all regions of the Earth's surface [11].

Jordan has a unique geographical location between the subhumid Mediterranean climate in the northwest and the arid desert climate in south and east, specifically its location along the Dead Sea Rift Valley, which is part of the largest East African Rift system, that includes tectonic activity to produce geothermal energy sources [14, 15]. Geothermal energy utilization in the Jordanian desert has been investigated by different researchers [16, 17], as the desert composes more than 70% of Jordan's area. In the northeast and southeast of Jordanian desert, shallow geothermal resources were found in which temperature gradient showed a promising potential for power generation despite various political, technical, and financial limitations [16, 17]. Jordan has good potential for geothermal energy utilization for recreational applications in the form of underground hot water such as wells and thermal springs, with temperatures ranging from 20 °C to 62 °C [1] distributed across different regions in the country. Abu-Hamatteh et al. [15] suggested potential power production in Jordan using geothermal binary power plants and Stirling engine.

Geothermal energy has diverse applications: power production, heating and cooling systems, desalination, greenhouse heating, and agriculture-related applications such as irrigation, livestock heating, and soil sterilization [18, 19]. It is reliable, clean, nonpolluting, and environmentally friendly, with no waste or by-product generation; also it can be used directly from the source [1, 19]. It is observed that geothermal power plants do not occupy much land; thus, it helps in protecting the natural environment [20]. As an example, fossil fuel power plants require 35 times more land compared to geothermal power plants [21]. A 20 MW geothermal binary power plant needs 1415 m^2/MW; also a 110 MW flash geothermal plant needs 1260 m^2/MW compared to a 25 MW wind farm that needs 16,000 m^2/MW of land [21]. In Jordan geothermal energy is used mainly for recreation and irrigation purposes [19], in which it is recommended to be utilized in greenhouse heating and fish farming applications [19]. The main objective of this chapter is to summarize the recent work that has been conducted on the geothermal energy status in Jordan.

The overall structure of this chapter is presented as follows:

First, part 2 provides an overview of various renewable energy sources and their applications. Part 3 presents the background and assessment of geothermal energy in Jordan beginning in 1965. In part 4, the applications of geothermal energy within Jordan are explored. Geothermal energy storage methods are discussed in part 5, followed by part 6 which examines the environmental impacts of utilizing geothermal energy for power production and space heating. Part 7 outlines the potential challenges related to the utilization of geothermal energy and the construction of geothermal power plants. Finally, part 8 concludes the chapter.

2.2 Renewable Energy Resources in Jordan

Jordan consumes annually 959 GWh of total renewable energy sources, including solar, wind, hydropower, biogas, and biofuel energy sources [2]. The total power production for different renewable energy sources for 4 years is summarized in Fig. 2.2.

Jordan has abundant sunlight along with its geographic location which makes it well-suited for solar energy power production. There are many photovoltaic (PV) and concentrated solar power (CSP) energy projects that are installed, ranging from utility-scale solar farms to distributed solar installations on building rooftops. The average insolation intensity on horizontal surface is 5–7 kWh/m^2/day which is one of the highest worldwide [1]. The annual sunshine duration is 2900 hours, which is sufficient to provide enough energy for space heating and cooling purposes [1]. Concentrated solar power production in the Middle East is estimated to 1266 GWh in 2021 based on IRENA 2023 report data. Figure 2.3 presents the total solar energy generation for different years (from 2013 to 2021). Total solar energy production in Jordan in 2021 is 3713 GWh compared to 51 GWh in 2015.

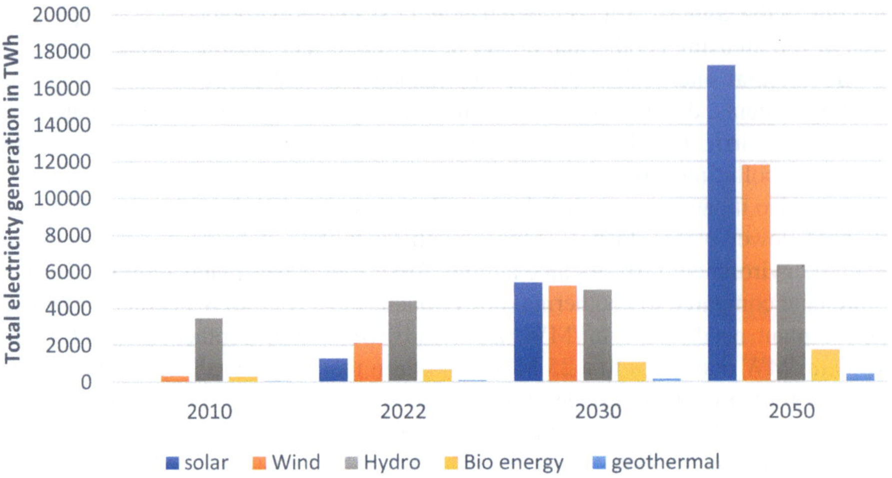

Fig. 2.2 Total power production in TWh for different renewable energy sources among different years (adapted from the International Renewable Energy Agency (IRENA) 2023 report)

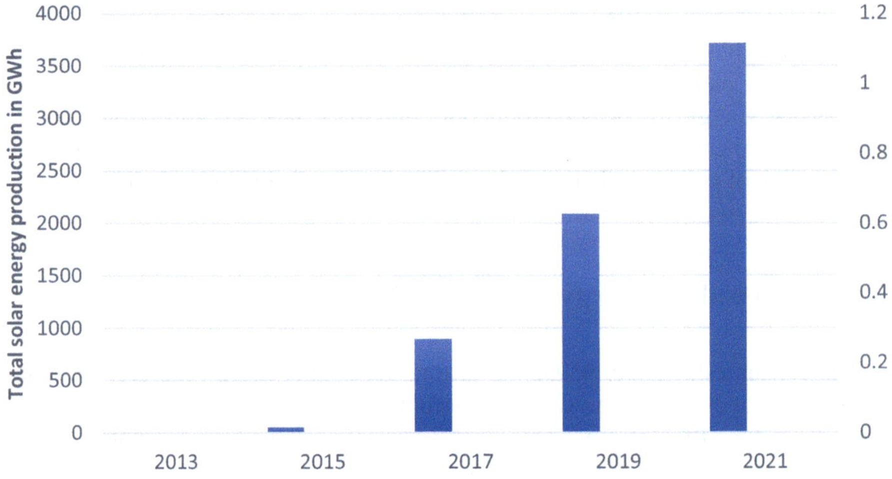

Fig. 2.3 Total solar energy production in Jordan in GWh among different years (adapted from the International Renewable Energy Agency (IRENA) 2023 report)

Another important renewable energy source in Jordan is wind energy. Many onshore and offshore wind projects are developed to increase the overall contribution of wind power to the grid. Different wind energy projects, particularly wind farms, are implemented to harness Jordan's wind resources. These projects involve the installation of wind turbines in locations with favorable wind conditions, such as Tafila, Fuhais, and Maán based on comprehensive wind assessments to ensure the economic viability and efficiency of power production [1, 22]. It is observed that the

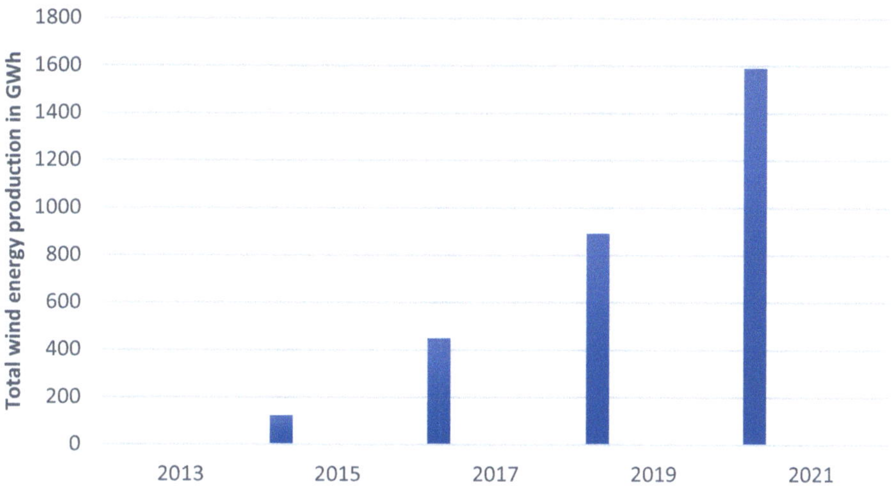

Fig. 2.4 Total wind energy production in Jordan in GWh among different years (adapted from the International Renewable Energy Agency (IRENA) 2023 report)

northern and southern parts of Jordan have high wind potential with an average wind speed of 6.5 m/s [1]. Figure 2.4 summarizes the total wind energy production for different years (from 2013 to 2021). The total wind energy production in Jordan in 2021 is 1590 GWh compared to 121 GWh in 2015.

On the other hand, hydropower energy sources in Jordan are very limited due to water shortage [23]. There are two main hydropower systems: the King Talal Dam, located on the Zarqa River near Jerash City, with an electricity generation capacity of 5 MW [24]. The second station is Aqaba Thermal Power Station of a turbine operating on the available head of cooling seawater boasting a capacity of 5 MW (Jordan Valley Authority, 2007), and the total electricity produced in 2007 was 45 GWh, constituting only 0.58% of the overall national power production. The total hydropower production for different years (from 2013 to 2021) is presented in Fig. 2.5. The total hydropower energy production in Jordan in 2021 is 18 GWh compared to 53 GWh in 2015. There is a significant reduction in utilizing hydropower energy for power production due to different limitations including the lack of both financial and technological resources to build its own traditional hydroelectric ability [25]. Furthermore, hydropower system relies on precipitation rates, which vary from year to year resulting in a hydroelectricity insecurity [24, 25].

For biomass resources in Jordan, they are utilized mainly for power and heat generation specifically in rural areas [26]. It was observed that Jordan produces large amount of solid waste as presented in Fig. 2.6 [27]. It is expected that the total municipal solid waste produced by households, commercial institutions, and industrial sectors will reach up to 3000 tons by 2030. Jordan produces more than 5.83 MT of different wastes and residues annually, in which 42% are utilized for biogas generation [26]. It was found that the municipal solid waste produced in the middle region of Jordan has a high potential for biogas and power production, followed by

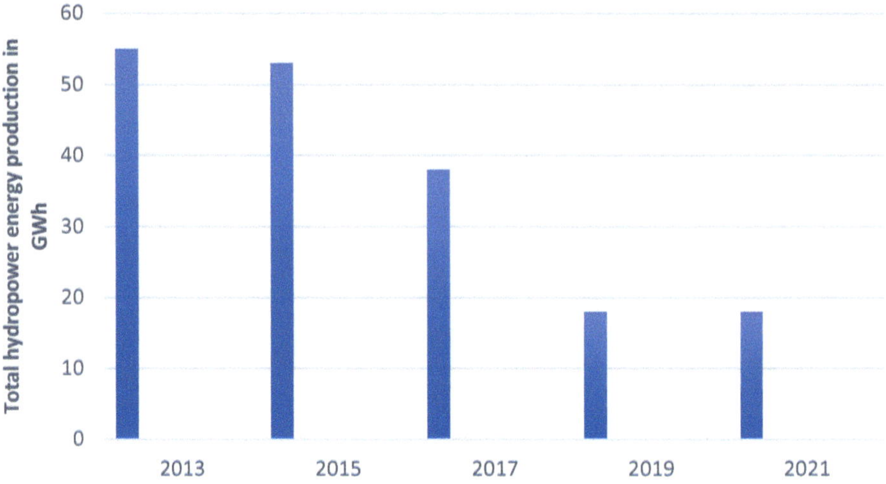

Fig. 2.5 Total hydropower energy production in Jordan in GWh among different years (adapted from the International Renewable Energy Agency (IRENA) 2023 report)

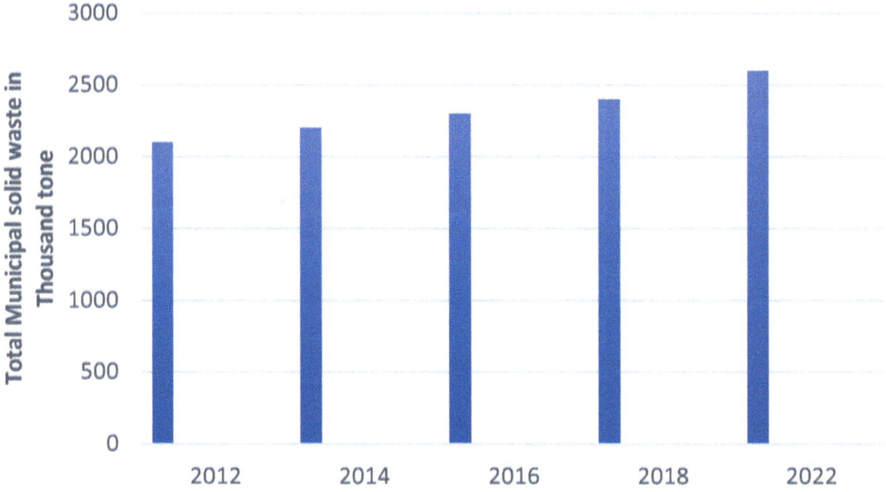

Fig. 2.6 Total solid waste production in Jordan (adapted from [26])

poultry manure and olive residues [26]. The total bio energy production in the Middle East and Jordan is estimated by 330 GWh and 4 GWh, respectively, in 2021 (adapted from IRENA 2023 report).

Geothermal energy sources have emerged as one of the most important renewable energy sources worldwide. Geothermal energy capacity around the world is 14,621 MW in 2021 compared to geothermal power production of 95,251 GWh in 2021 (based on IRENA 2023 report). The United States of America produces the highest geothermal capacity of 3900 MW compared to Indonesia in the second place of 2418 MW, while the Philippines in the third place of 1952 MW (adapted from ThinkGeoEnergy, 2023). In China as an example, geothermal energy in 2025 will be installed for more than 100 million m^2 of buildings (based on IRENA 2023 report). It is expected that geothermal energy power production will increase by 5% in 2050. Jordan has no electrical power generation from geothermal sources; it is mostly used for spas, recreation, and few agriculture activities. Therefore, it is important to develop a geothermal plan to explore and investigate geothermal sources due to their potential in different applications including direct thermal and power production applications [28, 29].

2.3 Geothermal Energy Background and Assessment in Jordan

In Jordan thermal springs and wells are mainly used for spa, recreation, and irrigation for many years [1, 15]. Most of thermal springs and hot boreholes in Jordan are located near the Dead Sea Rift area. There are many studies on geothermal energy status in Jordan that were carried out by the Natural Resources Authority (NRA) in collaboration with foreign institutions and companies. Over the years, various geothermal investigation campaigns have been carried out, with majority of the research focusing on the Zarqa Ma'in-Zara area. This region is characterized by deep groundwater circulation within normal geothermal gradient, resulting in temperature ranging between 30 and 63 °C.

Starting with 1965, MacDonald et al. [30] conducted chemical analysis of samples collected from Zarqa Ma'in area of Jordan. It was found that hydrochemical properties of thermal water was governed by the underlying geological formations and geothermal processes in the region. The analysis offered important insights into their mineral composition, temperature distribution, and the potential of geothermal energy exploitation in the area. It was found that thermal water in the Zarqa Ma'in area are characterized by elevated temperature and mineral content, which indicate a deep circulation of Earth's crust. Bender [31] in 1974 studied primary and secondary hot water spring locations focusing on the chemical analysis of thermal springs in the Zarqa Ma'in area, in which hydrological and geochemical characteristics of thermal springs were investigated. The findings contributed to map the geothermal potential in the region. McNitt in 1976 [32] investigated the existing chemical data from the Zarqa Ma'in area, and a resistivity survey was recommended along with drilling process to further study the area characteristics. Marinelli [33] in 1977 suggested that the Dead Sea Rift and the Zarqa Ma'in areas have the highest geothermal energy potential in Jordan, based on elevated temperature gradient and the chemical characteristics of thermal fluids in 1979, Truesdell [34] investigated the hydrogeochemical characteristics of geothermal spring and found that Zarqa Ma'in and Zara springs exist at a temperature of 110 °C and are cooled by mixing process between thermal and fresh cold water, in which the chemical compositing of spring water supported the mixing model. Furthermore, it was suggested that deep drilling and geophysical assessments are required to understand the behaviour of hotter, uncooled geothermal fluids. A geothermal reservoir was identified by Abu-Ajamieh in 1980 [35], which was heated by Hammamat Umm Hasan basalt plug as a primary heat source. His study confirmed the magmatic impact in Jordan's geothermal system that support the case of geothermal energy development in volcanic and rift-related settings. A follow up investigation on the previously mentioned studies was conducted by Mabey [36], who investigated further the geothermal energy status in the Zara-Ma'in area in order to refine the understanding of subsurface conditions, heat source, and geothermal potential. His findings resulted in various recommendations for the exploratory drilling and also confirmed that Zarqa-Ma'in area is a high priority site for geothermal energy development in Jordan. Hakki and Teimeh [37] in 1981 provided a comprehensive geological characteristics of the Zarqa Ma'in-Zara area, which was identified as one of the most important geothermal fields in Jordan. A detailed mapping of rock units focusing on lithostratigraphy of sedimentary and volcanic activities was conducted to understand the geological setting of the Zarqa Main area. Paola in the same year [38] provided that the temperature gradient variation in the thermal springs is due to the deep circulation of water in a typical geothermal field. This study concluded that Zarqa Ma'in area is governed by a combination of deep ground water circulation, structural control, and regional geothermal gradients not only shallow heating and localized anomalies. In 1982, McEwen and Holcombe [39] implemented terrain correction on the resistivity data presented by Abu-Ajamieh [35]. The adjusted data indicated the presence of four localized low-resistivity zones in the Zarqa Ma'in-Zara area, but none of these areas exhibited lateral continuity. In 1984, Flanigan and El-Kaysi [40] investigated the rock thermal conductivity characteristics in the Zarqa Ma'in area and found that it ranges from moderate to high thermal conductivity, which indicates that rocks are efficient to transfer heat required for geothermal energy exploitation. Galanis et al. [41] in 1986 studied the heat flow in the Zarqa Ma'in area and found that it can reach up to 472 mW/m^2, which is associated with fault zone related to the dead sea rift system. This study emphasized the importance of structural geology in geothermal system dynamics. Myslil in 1988 [42] investigated the heat flow data presented in the Galanis et al. [41]. Two zones were proposed based on temperature gradient that can reach to 50 °C/km, which provided a targeted approach for future exploration, drilling and geothermal energy planning. Allen in 1988 [43] summarized previous studies that were conducted in Jordan and proposed future work needed for effective exploration and utilization of geothermal energy. It was provided that geological and thermal conditions in parts of Jordan are highly favorable for geothermal development also a strong correlation between fault zone and geothermal activity was highlighted. Swarieh in 1992 [44] investigated geothermal potential for direct use applications near Queen Alia Airport area due to its proximity to the dead sea rift area and availability of boreholes and hydrochemical data. Swarieh and Massarweh [45–47] over different years investigated the geothermal potential in different locations in Jordan especially Karak and Zara-Ma'in areas, where they found that the stored heat is due to saturated sandstone which results in higher temperature gradient due to their high porosity and premeability allowing water to circulate and become thermally enriched. In 1997, thermal water in Mukhiebeh and Shuneh thermal well fields were investigated. In 1999, Saudi [48] investigated the thermal wells near Queen Alia Airport area and found that the maximum temperature exceeds 100 °C. In 2004, another study was conducted by Saudi et al. [49] on thermal wells located in central Jordan area. It was found that the highest temperature using mineral saturation index for thermal reservoir is 115 °C with high radon concentration located in fault intersections. Harayshat [1] estimated the geothermal energy capacity of 153.3 MWt along with the annual energy supply potential of 1540 TJ/year in the form of domestic hot water for direct use applications such as bathing and swimming with an overall capacity factor of 0.42. Al-Dabbas in 2011 [50] investigated the utilization of geothermal energy for space heating application; it was observed that the geoexchange systems saved around 70% in the heating mode, and 50% in the cooling mode compared to conventional fossil fuel systems.

Al-Zyoud et al. [51, 52] provided a geothermal source of a shallow aquifer system located in northeast Jordan for cooling applications, in which numerical 3D models were employed to forecast the future performance of geothermal reservior for potential development or installation. The study concluded that geothermal utilization of the investigated basaltic reservoir is feasibleas it exhibits adequate hydraulic and thermal characteristics suitable for cooling applications. Zaid et al. [19] carried out a comprehensive gravity survey in the Zarqa Ma'in and east of the Dead Sea areas. The findings were integrated with a range of previously acquired geophysical data, including aeromagnetic measurements, deep seismic soundings, and seismicity records. This combined analysis highlighted the structural and tectonics complexity of the region, revealing several local and regional Bouguer gravity anomalies of varying magnitudes, dimensions, and orientations. Theses anomalies are attributed to basic to ultrabasic igneous intrusions located within the uppermost crust, predominantly at depths ranging from 0.5–1.5 km or shallower [53]. Qasim in 2015 [54] proposed a binary power plant using hot geothermal water to generate electricity. Al-Hinti et al. [55] conducted an experimental study to establish the underground temperature distribution profile over one year near Zarqa city in Jordan. A simplified semi-empirical model was developed to predict the temperature profile as function of depth, time, the mean ground surface temperature, and soil properties, utilizing measured data to support the analysis. Aljaradin et al. in 2017 [56] compared the water quality of Afra Hot Springs with Dead Sea water. Slight variations between physicochemical parameters and trace metal contents were found between Afra Hot Spring and Dead Sea water due to surface runoff interactions. Afra Hot Spring water is considered as low-enthalpy geothermal sources utilized for recreational applications since the temperature ranges between 45 and 48 °C, in which power production is not possible due to low temperature gradient. Al-Zyoud in 2019 [16] suggested direct and indirect geothermal applications in Jordanian deserts. It was observed that heating, cooling, and crops drying are considered as the most common applications of geothermal energy in desert areas. In another study conducted by Al-Zyoud in 2019 [17] four geothermal potential fields in Jordan were evaluated, in which a geothermal gradient map was generated to present that Jordan has two high gradient fields with higher than 4.5 °C/100 m and other two moderate gradient fields with a range of 3.5–4.5 °C/100 m. In addition, it was found that the water temperature in some geothermal fields can reach to 68.5 °C. Alarmouty et al. in 2020 [57] investigated different case studies of residential and office buildings in Jordan for the potential of producing energy from near geothermal sources. It was found that geothermal power systems in buildings at a depth of 20 m are sufficient to collect the required energy for medium-sized residential building with an area around 300 m^2. Badran et al. in 2021 [58] suggested a binary cycle operated by a geothermal source for electrical power production of a health resort. It was suggested that the geothermal power plant might be an option for electrical power production of the Jordanian volcanic mountain resorts. Ayadi et al. in 2023 [59] conducted a comprehensive review of previous studies associated with local geothermal resources, focusing on the occurrence of theses sources and their potential for utilization.

Geothermal assessment is essential to investigate the potential application for different geothermal sources based on their temperature gradient. It is defined as an evaluation of the potential for utilizing geothermal energy at a specific location that includes investigating thermal, geological, and geophysical characteristics [60]. It is important to determine the feasibility and viability of geothermal energy projects, which helps in decision-making for geothermal development ensuring that such resources are being utilized in efficient and sustainable way [61]. Different geothermal assessment studies have been conducted in Jordan to continuously explore the geothermal energy potential for direct and indirect applications [1, 2, 5, 15–17] through geological, geophysical, temperature gradient measurements, hadrochemical analysis for groundwater and gas composition, and environmental impact assessment on biodiversity, land use, gas emissions, and water resources [62]. Geothermal assessment is also important to identify specific areas with high-temperature gradient that can be used for potential power production, in which the typical temperature gradient range between 100 and 180 °C [63]. Valgardur [64] assessed the geothermal energy status worldwide and investigated the geothermal energy stored quantity and quality that can be utilized by human beings. Different methods are used to estimate the production capacity for a given geothermal field including the volumetric assessment method and the geothermal simulation method. The volumetric method is most commonly used, which depend on thermal energy storage in a defined volume (e.g, rock and fluid) within a geothermal reservoir. The simulation method is conducted though developing a numerical model of the geothermal reservoir to simulate fluid flow, heat transfer and pressure behavior over time. It is observed that the estimates provided by the volumetric method are four to five times higher compared to simulation related methods.

Jordan has no power production from geothermal energy sources due to low to moderate temperature gradient, in which most of the geothermal energy in wells and thermal springs is used for recreational, spa, swimming pool heating, and agriculture-related activities. Potential areas in Jordan where geothermal energy can be utilized are presented in Fig. 2.7. Four main geothermal fields are presented, and

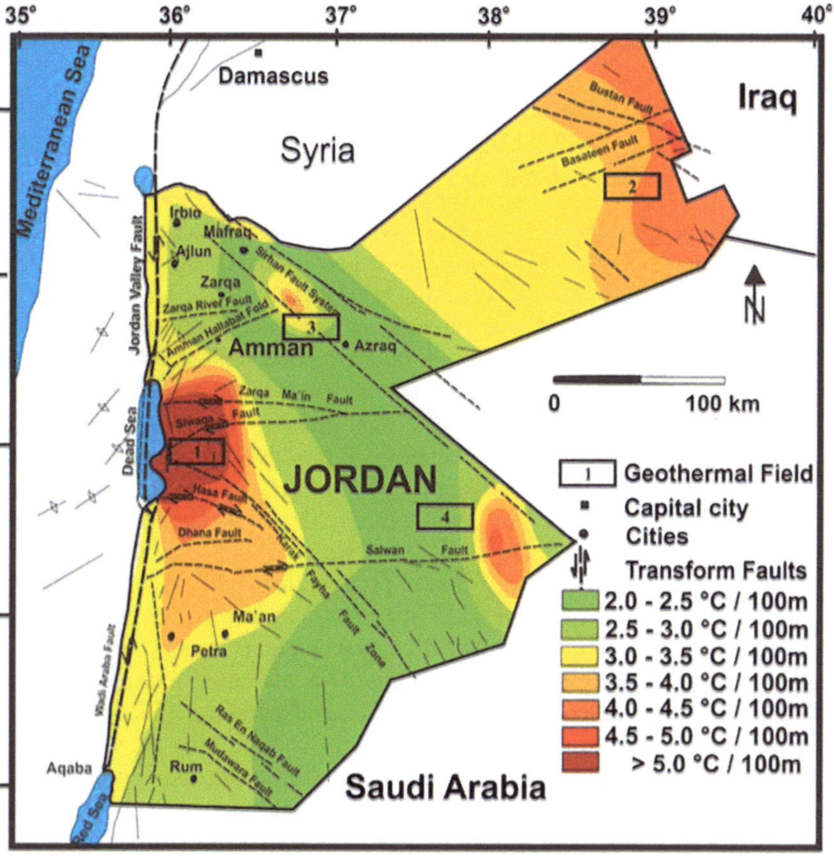

Fig. 2.7 Geothermal map of Jordan [17]

the temperature gradient distribution is shown. The fields are: Dead Sea Fault Geothermal (Field 1), Ar-Risha Geothermal (Field 2), Al Khalidiyya Geothermal (Field 3), and SE Desert Geothermal (Field 4) [17].

The geothermal gradient map (Fig. 2.7) reveals two areas with elevated temperature gradients, that can reach up to 50 °C/km [17]. The first is located near the East Dead Sea Escarpment where different thermal springs release heated water using lower Cretaceous sandstone. The second region is located near Syria and Iraq borders, where different thermal wells discharge water using upper Cretaceous limestone [65]. Thermal water from both thermal springs and wells exhibits temperatures ranging between 30 and 63 °C. The most popular area for wells and springs is the north Shuneh thermal well located in northwestern Jordan where the temperature gradient reaches 57 °C with 700 m^3/h flow rate compared to Afra Spring located in Tafilah south of Jordan with an average temperature of 47 °C and 376 m^3/h flow rate. Himmeh and Mukheibeh thermal well located in northern Jordan, the temperature gradient is between 38–44 °C and 30–41 °C, respectively with 300–3000 m^3/h and 200–6000 m^3/h flow rate, respectively. Smeika thermal well located in the Zarqa Main area has an average temperature of 57 °C with 50 m^3/h volume flow rate [15].

2.4 Geothermal Energy Applications in Jordan

Geothermal energy has been used for a wide range of applications worldwide [66]. One of the most important applications for geothermal energy is electrical power generation, space heating and cooling, greenhouse heating, and refrigeration for preserving food [5]. For power production applications, the Earth's internal heat is utilized to generate electricity through the extraction of steam from a geothermal reservoir to drive large-size turbines producing mechanical power; then turbines are connected to generators to produce electrical power [67]. Geothermal energy is a continuous and reliable power source; however, for power production applications, it was found that a significant amount of greenhouse gas (GHG) emissions are produced mostly in the form of carbon dioxide (CO_2) [68]. The global average CO_2 gas emission was estimated 122 g/kWh [68], which is still lower compared to traditional base load thermal energy power production facilities. Geothermal power plants consume 1022 L/MWh of water in a full life cycle for different power generation applications compared to coal fired power plants which consume approximately 2220 L/MWh [69]. Geothermal energy is also used for space heating and cooling in residential, commercial, and industrial buildings [57, 70]. Heat pump systems transfer heat from the ground to warm spaces during the winter and from warm spaces to ground during the summer season using the vapor compression refrigeration cycle [67]. They are preferred due to their high efficiency compared to conventional systems [55], and they offer better energy performance with lower energy costs. AlHinti et al. [55] investigated the temperature profile in Zarqa area and observed that underground temperature variations stabilize at 21 °C from 5 m below the earth surface, which indicated that it is feasible to use a ground source heat pump at shallow depth. In addition, Al-Ma'asfeh et al. [71] explored the potential of using geothermal energy for heating Al-Shaydia Phosphate Mining, which is based on utilizing hot water from the Shaydia mine to heat the building. Energy savings were estimated around $300,000 per year.

One of the largest geothermal energy projects that are utilized in Jordan for space heating and cooling is the ground source heat pump located in the American University of Madaba installed in 2012. It covers 1.6 MW cooling load and 1.5 MW heating load with 420 boreholes 100 m depth [5]. It was observed that the CO_2 emissions reduced by 47% compared to conventional systems such as chiller and LGP boiler systems. For the annual energy saving was estimated around 200, 000 kWh. Another geothermal project utilized in Jordan in 2017 is the Higher Council for Science and Technology (HCST), which covers 228 kW cooling load, 262 kW heating load, and domestic hot water for the building with 30 boreholes, each of which has dimensions (145 mm diameter, 130 m deep, and arranged in a 6 × 6 m grid). This system saved 40% of the cooling system's energy and 46% of the heating system's energy compared to conventional systems, which is about 19,784 JD with a payback period of 14 years [5].

Geothermal energy is utilized in greenhouse heating applications to maintain optimal temperatures for plant growth, extend the growing season, and efficient contribution to agricultural practices, which is useful in cold climates or during low ambient temperature periods [72, 73]. Emeish [74] proposed the utilization of geothermal energy for greenhouse heating in Jordan. Two methods were utilized for calculating the required heating load; static and dynamic methods. The static method calculates the heat loss due to infiltration and conduction through the greenhouse cover at the minimum ambient temperature, otherwise the dynamic method is more complex which incorporate mass and energy flow to the greenhouse while taking into account solar radiation, internal heat, and moisture dynamics. It was observed that static method utilized in greenhouse heating has 10% safety factor, which is preferred for proper accuracy. In Jordan, geothermal hot springs are very popular for their recreational and therapeutic purposes with more than 100 hot springs reported along the Dead Sea Rift area and the Zara–Zarqa Ma'in area [2]. Zarqa Ma'in area is very well known for its thermal springs, which utilize the natural geothermal heat for physical therapy and swimming pool heating. It is located near a plate boundary; thus, it is characterized by complex geology and tectonics [19]. These hot springs are low enthalpy, which means that the temperature gradient is low, thus, it cannot be used for efficient power production [75]. Jordan has conducted various studies to assess the potential for geothermal energy power production, which usually requires temperatures above 100 °C [76]. Other applications include water desalination, in which geothermal energy can be integrated into a desalination processes via RO technology that will help to provide energy for the separation of salt from seawater in order to produce fresh water where medium enthalpy geothermal fields with temperature range between 110–114 °C were discovered by the Jordanian Natural Resource Authority (JNRA) and British Geological Survey [2]. Aljaradin et al. [56] proposed the reuse of 3.3 million m^3 of thermal water extracted from hot springs to improve water availability in the Dead Sea region. The Afra spring, with water temperatures ranging between 45 and 48 °C, is classified as a low-enthalpy geothermal source. While it not suitable for power generation, it presents significant potential for development in therapeutic water tourism.

2.5 Geothermal Energy Storage

Energy storage is important for continuous power production; however, it has many limitations; geothermal energy is a continuous base load power source, where it provides a constant and reliable power output without any significant fluctuations compared to solar and wind energies [67, 77]. There are many indirect methods that can be utilized to store geothermal energy that involve the storage of heat generated by a geothermal source in a thermal energy storage (TES), which is considered as an affordable, environmentally friendly, and sustainable system that can be used to generate power or space heating. Hot water reservoirs are used as a thermal energy storage system with additional power system that can be considered as a supplemental output. Furthermore, molten salt system is utilized to store thermal energy due to their favorable characteristics that include low vapor pressure, non-toxic nature, low cost, flexibility, and high thermal stability [78]. Also, it has an efficient heat storage and can be used to produce steam for power production applications when demand is extremely high [78, 79]. Furthermore, carbon dioxide (CO_2) is utilized for energy storage due to its efficient thermal properties [80], in which CO_2 is initially injected into a geothermal layer for energy collection; then it is directed to an oil reservoir for geothermal energy storage. This process is efficient for large-scale geothermal energy storageand beneficial for carbon neutrality [80].

For geothermal energy storage systems, three methods are often discussed: Underground thermal energy storage (UTES), Aquifer thermal energy storage (ATES), and Borehole thermal energy storage (BTES] methods. The UTES method requires specific geological formations along with high initial costs. The ATES is referred to an open systems which offers large storage capacity and low operating costs; however, its applicability is restricted to regions with suitable aquifers, and it has the risk of aquifer contamination. For the BTES it is a closed system method, that requires minimum land footprint, and it is adaptable to different ground conditions [81].

A hybrid geothermal energy system combined with other forms of renewable energy, such as solar or wind energy, is proposed for more flexibility in energy production management and demand [65], in which excess energy from one source can compensate for fluctuations in another energy source. Morrone and Algieri [82] proposed a geothermal hybrid system utilized for a combined heat and power production for small-scale applications. When investigating energy storage methods, it is essential to consider geothermal conditions, local energy grid, and power demand to optimize the efficiency of energy production compared to large energy storage for extended period of time. In addition, geothermal energy can be used to compress air and store it in underground caverns [83], when electricity demand is high, the compressed air is released to drive turbines and produce electrical power.

2.6 Geothermal Environmental Impact

Geothermal energy has a lower environmental impact compared to fossil fuel-based energy sources, particularly in terms of reduced carbon emissions [21, 28]. However, environmental factors such as water contamination and air pollution must be carefully considered when utilizing geothermal energy for power production and space heating applications [20, 21, 84]. Geothermal power plants have a factor of 10 times lower carbon dioxide (CO_2) gas emissions compared to gas-fired power plants and a factor of 20 times lower compared to oil and coal-power generation plants [21]. It produces one-sixth of CO_2 compared to natural gas power plants [28]. Geothermal closed-loop power plants have zero emissions of CO_2, SO_2, and NOx compared to conventional power plants. Geothermal power plants release significant amounts of heat into the environment, whether into the air or a nearby water source, which will have a significant impact on biodiversity. It has been observed that the highest levels of heat rejection occur in binary cycle geothermal power plants, while the lowest levels were associated with steam power plants [28]. For land use, geothermal power plants need less land compared to other power plants. For example, a 20 MW geothermal binary power plant requires 1415 m^2/MW of land compared to a 25 MW wind farm that requires 16,000 m^2/MW of land, and a 2300 MW coal power plant requires about 40,000 m^2/MW of land [21]. Generally, fossil fuel power plants require 35 times more land compared to geothermal power plants. For water consumption, geothermal power plants consume an average of 1022 l/MWh, whereas coal-fired power plants use 2220 l/MWh [69]. The total water consumption required to drill a 1-m well ranges from 5 m^3 to 30 m^3, depending on the technology employed, geological conditions, and number and depth of liners being used [85].

Geothermal power plants produce solid waste from cooling water scaling, drill cutting, construction waste, maintenance waste, silica sludge, organic waste, and machine and heavy equipment waste [86]. As an example, in Wayang Windu Geothermal Field in Indonesia in 2017, the total amount for hazardous and toxic waste was approximatly 10.165 ton compared to 10.727 ton in 2016, in which the waste was reduced by 5% after applying different engineering control measures and soil waste management strategies. On the same geothermal field, the organic waste for 2016 and 2017 was approximately 12,696 and 12,708 tons, respectively, which can be eventually compostable.

2.7 Geothermal Energy limitations

There are many challenges associated with geothermal energy utilization, such as geological structures that require extensive exploration and exploitation methods to be utilized to access existing resources. Furthermore, technological challenges that can be found in geothermal wells drilling, limited knowledge for precise availability of geothermal resources [87], limited transportation of heavy equipment utilized in geothermal exploration and development phase, limited economic competitiveness of geothermal power, risk associated with acquiring geothermal resources at an economically viable rate, and soil contamination which will have a significant impact on the environment including air, water, and solid waste pollution [84]. Geothermal energy in Jordan mainly is utilized for direct applications including space heating and cooling, which involves high initial installation costs, especially the drilling cost [88, 89]. It is observed that the total power produced from a geothermal energy source is considered small compared to other renewable sources such as solar and wind energies [19]; therefore, geothermal energy is not used on a large-scale system for power production worldwide, because it takes long period of time from when the resource is discovered until when the power plant is being fully functional [29]. It is very important to conduct feasibility analysis prior to any geothermal power plant installation, which will justify the capital expenditure and operational costs [29].

Hu et al. [87] discussed the geothermal power production and direct use applications in China. Power production was limited by insufficient thermal gradient close to major demand areas. Many challenges were described including drilling under extremely high-temperature and high-pressure conditions, long-distance electrical transmission, heavy equipment transportation, and the exploitation of hot dry rock (HDR) resources. Their paper recommended the investigation of enhanced geothermal system (EGS) technologies in order to increase the growth of geothermal energy market especially in dense population areas [90]. Kumar et al. [91] discussed the challenges associated with utilizing geothermal energy for direct use applications such as space heating and cooling. Their study provided that one of the primary challenges in developing geothermal projects is the limited availability of comprehensive data on potential geothermal energy resources.

2.8 Conclusion

Geothermal energy is an important renewable energy source utilized worldwide for different direct and indirect applications. It plays a vital role in providing clean and reliable energy sources, and it is effective in reducing greenhouse gas emissions along with global warming effects. Despite the availability of some geothermal energy sources in Jordan, these sources are not fully utilized to develop and enhance the Jordanian energy sector, which are mainly used for space heating, cooling, recreation, and agricultural activity applications. There are four main geothermal fields in Jordan in which the temperature gradient is low, means that geothermal energy cannot be utilized for power production. However, it is recommended to further explore the feasibility of geothermal recourses in Jordan, which involve conducting further exploration and assessment studies to accurately identify and evaluate different areas for geothermal energy potential either for power production or further direct use applications. Finally, it is very important to continuously invite the private sector to invest and participate further in the geothermal energy investigation and various exploration campaigns.

References

1. E.S. Hrayshat, Status and outlook of geothermal energy in Jordan. Energy Sustain. Dev. **13**(2), 124–128 (2009)
2. P. Amoatey, M. Chen, A. Al-Maktoumi, A. Izady, M.S. Baawain, A review of geothermal energy status and potentials in middle-east countries. Arab. J. Geosci. **14**, 1–19 (2021)
3. A. Raihan, Dynamic Impacts of Economic Growth, Renewable Energy, Natural Resources, and Globalization toward Environmental Sustainability in Jordan, in *Proceedings of the International Conference on Climate Change and Natural Resources Management for Sustainable Development*, (Springer, Cham, 2024)

4. A.A. Rafindadi, Z. Yusof, K. Zaman, P. Kyophilavong, G. Akhmat, The relationship between air pollution, fossil fuel energy consumption, and water resources in the panel of selected Asia-Pacific countries. Environ. Sci. Pollut. Res. **21**, 11395–11400 (2014)
5. Y.M. Al-Smadi, A.M. Alshorman, W. Hassan, R. Bader, I.A. Awad, S. Alzghoul, H. Bataineh, Assessment and perception of renewable energy awareness and potential in Jordan. Jordan J. Mech. Ind. Eng. **16**(4), 615–625 (2022)
6. J.S. Riti, Y. Shu, Renewable energy, energy efficiency, and eco-friendly environment (RE 5) in Nigeria. Energy Sustain. Soc. **6**, 1–16 (2016)
7. I. Stober, K. Bucher, *Geothermal Energy* (Springer, Berlin, 2013)
8. K. Silaipillayarputhur, T.A. Mughanam, Utilization of geothermal energy in King Faisal University's farm. AIP Conf. Proc. **2933**(1), 020003 (2023)
9. Z. Yusupov, M. Almaktar, Geothermal Power Generation, in *Geothermal Energy*, (IntechOpen, London, 2021)
10. S.B.A. Kashem, M. Hasan-Zia, M. Nashbat, A. Kunju, A. Esmaeili, A. Ashraf, M.E. Chowdhury, A review and feasibility study of geothermal energy in Indonesia. Int. J. Technol. **2**(1), 19–34 (2021)
11. R. Guerrero-Lemus, L.E. Shephard, R. Guerrero-Lemus, L.E. Shephard, Geothermal Energy, in *Low-Carbon Energy in Africa and Latin America: Renewable Technologies. Natural Gas and Nuclear Energy*, (Springer, Cham, 2017), pp. 243–259
12. M.H. Dickson, M. Fanelli, What is Geothermal Energy? in *Renewable Energy*, (Routledge, London, 2018), pp. 302–328
13. H. Muraoka, Geothermal Energy, in *Handbook of Climate Change Mitigation and Adaptation*, (Springer International Publishing, Cham, 2022), pp. 809–836
14. P. Omenda, M. Teklemariam, Overview of Geothermal Resource Utilization in the East African Rift System, in *Short Course V on Exploration for Geothermal Resources*, (Geological Survey of Ethiopia, Addis Ababa, 2010)
15. Z. Abu-Hamatteh, K. Al-Zughoul, S. Al-Jufout, Potential geothermal energy utilization in Jordan: Possible electrical power generation. Int. J. Thermal Environ. Eng. **3**, 9–14 (2011)
16. S. Al-Zyoud, Geothermal energy utilization in Jordanian deserts. Int. J. Geosci. **10**(10), 906–918 (2019)
17. S. Al-Zyoud, Shallow geothermal energy resources for future utilization in Jordan. Open J. Geol. **9**(11), 783 (2019)
18. J. Bundschuh, G. Chen, D. Chandrasekharam, J. Piechocki (eds.), *Geothermal, Wind and Solar Energy Applications in Agriculture and Aquaculture* (Routledge, London, 2017)
19. A.I. Zaid, H. Zuhair, R.J. El-Kilani, Utilization of Geothermal Energy in Jordan, in *2015 Power Generation System and Renewable Energy Technologies (PGSRET)*, (IEEE, Piscataway, 2015), pp. 1–9
20. R. DiPippo, *Geothermal Power Plants: Principles, Applications, Case Studies and Environmental Impact* (Butterworth-Heinemann, Oxford, 2012)
21. M. Soltani, F.M. Kashkooli, M. Souri, B. Rafiei, M. Jabarifar, K. Gharali, J.S. Nathwani, Environmental, economic, and social impacts of geothermal energy systems. Renew. Sust. Energ. Rev. **140**, 110750 (2021)
22. S.M. Habali, M. Amr, I. Saleh, Ta'ani, R., Wind as an alternative source of energy in Jordan. Energy Convers. Manag. **42**(3), 339–357 (2001)
23. N. Hadadin, M. Qaqish, E. Akawwi, A. Bdour, Water shortage in Jordan—Sustainable solutions. Desalination **250**(1), 197–202 (2010)
24. W.S. Maisa'a, M.M. Abualhaija, An analysis of long term yearly water flow trend and its impact on sediment yield in king Talal dam. Int. J. Eng. Res. Technol. **12**, 3041–3049 (2020)
25. C.S. Kaunda, C.Z. Kimambo, T.K. Nielsen, Hydropower in the context of sustainable energy supply: A review of technologies and challenges. Int. Sch. Res. Notices **1**, 730631 (2012)
26. Z. Al-Hamamre, M. Saidan, M. Hararah, K. Rawajfeh, H.E. Alkhasawneh, M. Al-Shannag, Wastes and biomass materials as sustainable-renewable energy resources for Jordan. Renew. Sust. Energ. Rev. **67**, 295–314 (2017)
27. A.A.K. Al-Bawwat, F. Jurado, M.R. Gomaa, A. Cano, Availability and the possibility of employing wastes and biomass materials energy in Jordan. Sustain. For. **15**(7), 5879 (2023)

28. L.D. Berrizbeitia, Environmental Impacts of Geothermal Energy Generation and Utilization. Volcanos of the Eastern Sierra Nevada–G190. Hamburger, Rupp and Taranovic, 2014
29. M.J.B. Kabeyi, Geothermal electricity generation, challenges, opportunities and recommendations. Int. J. Adv. Sci. Res. Eng. **5**(8), 53–95 (2019)
30. M. MacDonald, Report to central water authority, Hashemite Kingdom of Jordan, 1965
31. F. Bender, *Geology of Jordan. Contribution of the Regional Geology of the Earth* (Gebruder Borntraeger, Berlin, 1974)
32. S. McNitt, Geothermal resources in Jordan - interpretation of data submitted by the government. UNDP report, 1976
33. G. Marinelli, Report on the Possibility of Developing Geothermal Resources in Jordan. University of Pisa, report, 1977
34. A.H. Truesdell, *Final Report on the Chemistry and the Geothermal Energy Possibilities of Zarqa ma'in—Zara Hot Springs, Jordan* (U.S. Geol. Survey, Menlo Park, 1979)
35. M. Abu Ajameih, The geothermal resources of Zarqa ma'in and Zara. Phase 1 of the geothermal energy project in Jordan. Natural Resources Authority, Amman, report (1980), p. 82
36. R.D. Mabey, Recommended program for the Zarqa ma'in and Zara area. U.S. Geol. Survey, report, 1980
37. W. Hakki, M. Teimeh, The Geology of Zarqa ma'in and Zara Areas. NRA, Amman, report, 1981
38. G. Di Paola, *Mission Report on Visit to the Geothermal Project JOR/76/004 in the Hashemite Kingdom of Jordan from 30 November to 4 December 1981* (UN/DTCD, New York, 1981)
39. R. McEwen, H. Holconmbe, Interpretation of Resistivity Data in Zarqa ma'in and Zara Hot Springs. Exploration Geothermics. Santiago, California, USA, 1982
40. E.L. Flanigan, Q. Kaysi, Preliminary interpretation of audio-Magnetotelluric (ATM) ounding in the Zara-Zarqa Ma'in geothermal areas Kingdom of Jordan. Preliminary report USGS, 1984
41. S.P. Galanis, J.H. Sass, R.J. Munroe, M. Abu Ajameih, Heat flow at Zarqa Ma'in and Zara and a geothermal reconnaissance of Jordan. U.S. Geological Survey, open-file report, 86-631,110 pp, 1986
42. V. Myslil, Report on evaluation of geothermal potential of Jordan. Strojexport, Prague, Geological Survey, Prague, report, 1988
43. D. Allen, *Preliminary Evaluation of the Geothermal Potential of Jordan Recommendations for Future Studies* (BGS, Keyworth, 1988), p. 34
44. A. Swarieh, Thermal Boreholes near Queen Alia Airport. NRA, Amman. 47 p, 1992
45. A. Swarieh, R. Massarweh, Thermal springs in Wadi Ibn Hammad. NRA, Amman. 32 p, 1993
46. A. Swarieh, R. Massarweh, Geothermal Water in Zara and Zarqa ma'in Area. Report No. 8. NRA, Amman. 36 p, 1995
47. A. Swarieh, R. Massarweh, Geothermal water in Mukhiebeh and North Shuneh areas. Report No. 9. NRA, Amman. 26 p, 1997
48. A. Saudi, The geochemistry of thermal fluid in the geothermal field near alia airport in Jordan and Selfoss geothermal field, S-Iceland, Reports of the United Nations University, report no. 13. 354 p, 1999
49. A. Saudi, A. Swarieh, A. Diabat, A. Masri, R. Masarwih, Thermal Fluids in Central Jordan, NRA Amman. 125 p, 2004
50. M.A.A. Al-Dabbas, Achievement of geothermal energy using ground heat exchanger in ma'en. J. Mech. Sci. Technol. **25**, 2013–2023 (2011)
51. S. Al-Zyoud, W. Rühaak, I. Sass, The potential of shallow groundwater resources for cooling purposes-a geothermal case study in North East Jordan, 2012
52. S. Al-Zyoud, W. Rühaak, I. Sass, Dynamic numerical modeling of the usage of groundwater for cooling in north East Jordan–a geothermal case study. Renew. Energy **62**, 63–72 (2014)
53. C. Chisenga, Understanding the Earth Structure underneath Botswana: The Tectonic Model and its Relationship to the Basement and Crustal Thickness, Master's thesis, University of Twente, 2015
54. F. A. Qasim Geothermal Power of a Hot Spring in Jordan, Master's thesis, University of Massachusetts Lowell, 2015

55. I. Al-Hinti, A. Al-Muhtady, W. Al-Kouz, Measurement and modelling of the ground temperature profile in Zarqa, Jordan for geothermal heat pump applications. Appl. Therm. Eng. **123**, 131–137 (2017)
56. M. Aljaradin, R. Bashitialshaaer, H. Alitawi, M. Amaireh, Water quality, availability and potential of geothermal energy utilization, Afra water, Jordan. Sustain. Resour. Manage. J. **1**(1), 17–25 (2017)
57. T. Alarmouty, Q. Jobran, F. Sharaf, The effective of geothermal energy in buildings: a case study in Madaba City in Jordan, 2020
58. O.O. Badran, G.S. Al-Marahleh, A.F.O. AlKhawaldeh, I.A. Aldabaibeh, Design of a small capacity geothermal power plant for cooling and heating systems of health resort in Jordan. Curr. J. Appl. Sci. Technol. **40**(4), 14–29 (2021)
59. O. Ayadi, J.O. Jaber, S. Al-Zyoud, Geothermal energy resources in Jordan, 2023
60. J. Limberger, T. Boxem, M. Pluymaekers, D. Bruhn, A. Manzella, P. Calcagno, J.D. van Wees, Geothermal energy in deep aquifers: A global assessment of the resource base for direct heat utilization. Renew. Sust. Energ. Rev. **82**, 961–975 (2018)
61. R. Shortall, B. Davidsdottir, G. Axelsson, Geothermal energy for sustainable development: A review of sustainability impacts and assessment frameworks. Renew. Sust. Energ. Rev. **44**, 391–406 (2015)
62. I.K. Rotich, H. Chepkirui, P.K. Musyimi, G. Kipruto, Geothermal energy in Kenya: Evaluating health impacts and environmental challenges. Energy Sustain. Dev. **82**, 101522 (2024)
63. D. Moya, C. Aldás, P. Kaparaju, Geothermal energy: Power plant technology and direct heat applications. Renew. Sust. Energ. Rev. **94**, 889–901 (2018)
64. I.B. Fridleifsson, Status of geothermal energy amongst the world's energy sources. Geothermics **32**(4-6), 379–388 (2003)
65. M.A. Bagherian, K. Mehranzamir, A comprehensive review on renewable energy integration for combined heat and power production. Energy Convers. Manag. **224**, 113454 (2020)
66. G. Allegretti, M.A. Montoya, E. Talamini, Renewable Energy for a Sustainable Future, in *The Renewable Energy-Water-Environment Nexus*, (Elsevier, Amsterdam, 2024), pp. 1–36
67. C.I. Igwe, Geothermal energy: A review. Int. J. Eng. Res. Technol. **10**(3), 655–661 (2021)
68. T. Fridriksson, A.M. Merino, A.Y. Orucu, P. Audinet, Greenhouse Gas Emissions from Geothermal Power Production, in *Proceedings, 42nd Workshop on Geothermal Reservoir Engineering*, (Stanford University, Stanford, 2017)
69. Y. Jin, P. Behrens, A. Tukker, L. Scherer, Water use of electricity technologies: A global meta-analysis. Renew. Sust. Energ. Rev. **115**, 109391 (2019)
70. H. Hristov, N. Nikolova, K. Bojadgieva, Utilization of Geothermal Waters for Space Heating in Bulgaria, in *Proceedings of the World Geothermal Congress, Kyushu–Tohoku, Japan*, vol. 28, (International Geothermal Association, Tokyo, 2000)
71. S.I. Al-Ma'asfeh, M.R. Gomaa, A. Azetawi, A. Shwakih, M. Mehdawi, Exploiting geothermal energy as a source of heating system in Al-Shaydia phosphate mining: overview and case study, 2024
72. O. Ozgener, G. Kocer, Geothermal heating applications. Energy Sources **26**(4), 353–360 (2004)
73. F. Fahmy, D. Atia, H. El Madany, H. Farghally, Greenhouse heating systems based on geothermal energy. Technology **6**, 7 (2018)
74. M.S. Emeish, *Geothermal Heating System for Jordanian Greenhouses* (United Nations University, Shibuya, 1999)
75. A. Al-Sarkhi, B. Akash, E. Abu-Nada, S. Nijmeh, I. Al-Hinti, Prospects of geothermal energy utilization in Jordan. Energy Sour. Part A **30**(17), 1619–1627 (2008)
76. A. Manzella, Geothermal energy. EPJ Web Conf. **148**, 00012 (2017)
77. H. Kulasekara, V. Seynulabdeen, A Review of Geothermal Energy for Future Power Generation, in *2019 5th International Conference on Advances in Electrical Engineering (ICAEE)*, (IEEE, Piscataway, 2019), pp. 223–228
78. P. Bhatnagar, S. Siddiqui, I. Sreedhar, R. Parameshwaran, Molten salts: Potential candidates for thermal energy storage applications. Int. J. Energy Res. **46**(13), 17755–17785 (2022)

79. R. Roper, M. Harkema, P. Sabharwall, C. Riddle, B. Chisholm, B. Day, P. Marotta, Molten salt for advanced energy applications: A review. Ann. Nucl. Energy **169**, 108924 (2022)
80. Y. Liu, T. Hu, Z. Rui, Z. Zhang, K. Du, T. Yang, A. Afanasyev, An integrated framework for geothermal energy storage with CO2 sequestration and utilization. Engineering **30**, 121–130 (2023)
81. M. Shah, M. Prajapati, K. Yadav, A. Sircar, A comprehensive review of geothermal energy storage: Methods and applications. J. Energy Storage **98**, 113019 (2024)
82. P. Morrone, A. Algieri, Integrated geothermal energy systems for small-scale combined heat and power production: Energy and economic investigation. Appl. Sci. **10**(19), 6639 (2020)
83. J. Menéndez, A. Ordóñez, R. Álvarez, J. Loredo, Energy from closed mines: Underground energy storage and geothermal applications. Renew. Sust. Energ. Rev. **108**, 498–512 (2019)
84. W.E. Glassley, *Geothermal Energy: Renewable Energy and the Environment* (CRC Press, Boca Raton, 2014)
85. M. Bošnjaković, M. Stojkov, M. Jurjević, Environmental impact of geothermal power plants. Tehnički vjesnik **26**(5), 1515–1522 (2019)
86. A. Utami, N. Aji, A. Fadyah, A. Ghifari, M.B. Anam, S. Ramadhani, R.R. Maulana, Geothermal energy solid waste management: Source, type of waste, and the management. AIP Conf. Proc. **2245**(1), 060001 (2020)
87. Y. Hu, H. Cheng, S. Tao, Opportunity and challenges in large-scale geothermal energy exploitation in China. Crit. Rev. Environ. Sci. Technol. **52**(21), 3813–3834 (2022)
88. Z.G. Acevedo, M.S. Vázquez, J.G. Troncoso, S.L. Sánchez, M.G. Zarate, Social Acceptance/ Resistance of Geothermal Power Plants in Mexico, in *First EAGE Workshop on Geothermal Energy in Latin America*, (European Association of Geoscientists & Engineers, Utrecht, 2021), pp. 1–5
89. R. Bertani, Deep Geothermal Energy for Heating and Cooling, in *Renewable Heating and Cooling*, (Woodhead Publishing, Sawston, 2016), pp. 67–88
90. J.W. Lund, Direct utilization of geothermal energy. Energies **3**(8), 1443–1471 (2010)
91. Kumar, L., Hossain, M. S., Assad, M. E. H., Manoo, M. U. Technological advancements and challenges of geothermal energy systems: *a comprehensive review. Energies, 15*(23), 9058 (2022).

Chapter 3
Advancing Energy Efficiency in Smart Cities: The Smart Readiness Indicator (SRI) Framework for Sustainable Building Performance

Armin Bagherian, Mohsen Gharehdaghi, Mohammad Mohsen Hayati, Farbod Borhani, Mehdi Abapour, and Behnam Mohammadi-ivatloo

3.1 Introduction

Increasing global environmental concerns pose a significant challenge worldwide [1]. Also, access to sustainable energy sources has been a longstanding issue for developed and industrialized nations, with the European Union (EU) standing out as the world's largest energy consumption market [2, 3]. In recent years, the EU has implemented energy directives aimed at improving energy performance and sustainability in the built environment, with a particular focus on integrating smart technologies [4, 5]. Furthermore, there has been a growing emphasis on investing in smart, energy-efficient technologies for smart homes [6]. The EU emphasizes the pivotal role of the building sector in advancing its climate actions and long-term strategies towards achieving the energy transition goals by 2050 [7, 8]. From an energy viewpoint, smart structures such as homes and buildings must prioritize energy efficiency, meeting a significant portion of their energy needs through on-site renewable energy sources (RESs) to achieve nearly zero-energy building status

A. Bagherian (✉) · M. Gharehdaghi · M. M. Hayati · M. Abapour
Faculty of Electrical and Computer Engineering, Energy Systems Research Institute (ESRI), Smart Energy Systems Research Lab (SeS Lab.), University of Tabriz, Tabriz, Iran
e-mail: arminbagherian@ieee.org; mohammadmohsen.hayati@tabrizu.ac.ir; abapour@tabrizu.ac.ir

F. Borhani
Faculty of Computer Engineering and Information Technology, Amirkabir University of Technology, Tehran, Iran
e-mail: farbod.borhani@aut.ac.ir

B. Mohammadi-ivatloo
School of Energy Systems, Department of Electrical Engineering, LUT University of Technology, Lappeenranta, South Karelia, Finland
e-mail: behnam.ivatloo@lut.fi

A. Ahmadian et al. (eds.), *Sustainable Energy Resources in Smart Cities*, Green Energy and Technology, https://doi.org/10.1007/978-3-031-91516-1_3

[9, 10]. Approximately half of the final energy utilized in the EU is dedicated to heating and cooling according to the energy performance building directive (EPBD), with buildings accounting for 80% of this consumption [11]. The Buildings Performance Institute Europe (BPIE) characterizes a smart building as a dynamic micro-energy hub [12, 13]. These buildings are equipped with energy generation, control, storage, demand response capabilities, and can interface with electric vehicles [14, 15]. In other words, a smart building or home is described as a structure that is seamlessly integrated into a smart city system, actively contributing to energy storage, generation, and load management [16, 17]. This integration enhances energy system flexibility, welfare, and convenience of occupants while reducing energy consumption and carbon emissions [18]. The worldwide interest in "smart" and "green" buildings is rapidly growing due to their potential to enhance resource efficiency, promote renewable energy utilization, and contribute to mitigating climate change impacts [19]. In the EU, buildings are accountable for 40% of total energy consumption and 36% of greenhouse gas (GHG) emissions [20]. Globally, buildings consume 36% of total energy and emit approximately 37% of energy-related carbon dioxide [21]. The integration of digitalization and advancements in information and communication technology is crucial for enhancing the efficiency of the European energy market and advancing sustainable, renewable energy systems [22, 23]. Consequently, European Union Member States are urged to establish an optional common framework for defining and calculating a metric, known as a smart readiness indicator (SRI) [24, 25]. This indicator assesses buildings' capabilities to adjust their operations to meet occupants' needs and the electricity grid's requirements, promoting more efficient operation. Additionally, SRI has been identified as a valuable tool for building development endeavors, contingent upon accurate interpretation of results and proper execution of additional assessments. The calculation of this indicator employs a methodology developed by the European Commission Directorate-General for Energy, which relies heavily on factors such as building type and climatic conditions. The impact of these parameters on the SRI is demonstrated through weighting coefficients, which should be initially determined by legislative bodies. Therefore, the critical step in evaluating the effectiveness of this methodology is to test it across various scenarios, encompassing different building types, climatic conditions, and other relevant factors [26, 27].

The main contributions of this work are included:

(a) Standardized selection process: The SRI technique offers a standardized framework for selecting appropriate services based on the specific context of the building, climate, and site-specific factors.
(b) Performance criteria: The SRI technique considers various impact categories such as energy conservation, energy flexibility and storage, comfort, convenience, health, well-being, accessibility, maintenance, fault prediction, and information dissemination to occupants.
(c) Flexibility and adaptability: The SRI technique is adaptable and modular, allowing for flexibility to address local and site-specific contexts, ensuring that irrelevant or undesirable services are not included in the assessment.

(d) Enhanced energy efficiency: By evaluating smart-ready services and their impact on energy performance, the SRI contributes to improving energy efficiency in buildings.

3.2 The Energy Performance Buildings Directive (EPBD)

The EPBD, approved by the European Parliament and the Council, relates to the energy efficiency of buildings taken from the Kyoto Protocol. It includes EPBD (EU/2010/31) [28] and the Energy Efficiency Directive (EED) (EU/2023/1791) [29]. These protocols were revised in 2018 and 2023, and other laws were added. The importance of these directives is to promote highly efficient and decarbonized buildings by 2050 so that consumers and investors can make more informed decisions about energy and money for buildings.

On December 7, 2023, co-legislators (Council of the EU and the European Parliament) reached an interim agreement on EPBD revisions, which is formally adopted in 2024 [30]. Enhancing the energy efficiency of buildings yields substantial energy savings and reduced energy expenditures, thus mitigating energy poverty [31] and fostering greater energy self-sufficiency in Europe. Additionally, this endeavor contributes to the enhancement of citizens' health and well-being by elevating living standards to meet the demands of the twenty-first century for all individuals. Moreover, allocating resources towards energy efficiency initiatives serves to invigorate the energy economy and generate additional environmentally friendly employment opportunities [32]. Figure 3.1 shows the evolution of EPBD law over the years.

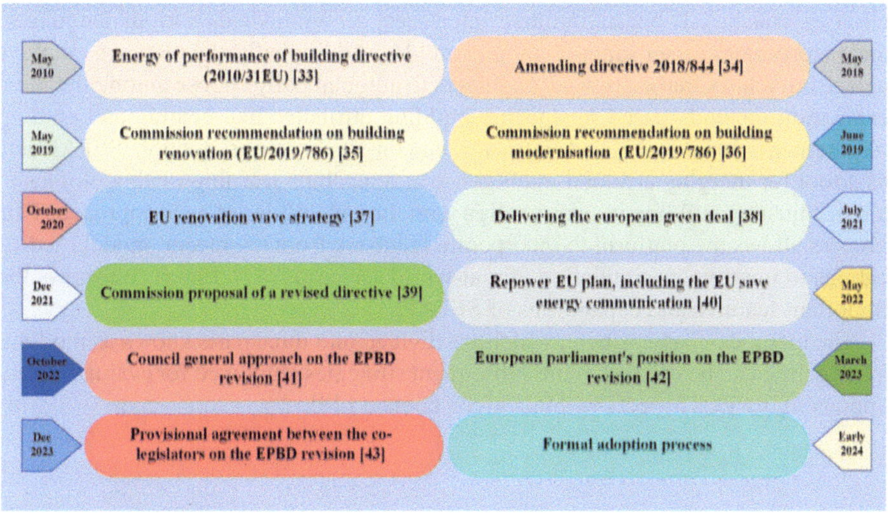

Fig. 3.1 EPBD timeline [33–43]

3.2.1 Energy Performance Assessment

The energy performance certificate (EPC) is a method used to evaluate a building's energy usage and ascertain its suitability. It offers insights into a building's energy efficiency by assigning it a rating ranging from A to G, reflecting its energy performance level. This certification aids property owners and inhabitants in comprehending the energy efficiency status of the building and offers suggestions for enhancing its energy efficiency [44]. EU member states established EPC systems in response to Directive 2002/91/EC. These systems aim to provide precise energy performance data for buildings, as well as guidance on improving performance and conducting cost-effective analyses of enhancement measures. Nevertheless, a decade following the enactment of the initial EPBD, the attainment of these goals remains incomplete [45]. Hence, there is a necessity to formulate and create an advanced iteration of EPC that is more dependable, user-centric, and economical. This is crucial to instill confidence in the market and encourage financial backing for energy-efficient construction. Presently, the utilization of the EPC is infrequent. It is commonly stowed away as a compulsory record due to its standardized nature. A beneficial, enhanced, and notably user-centric format for the energy performance certificate has been introduced [46].

3.2.2 The Smart Energy Performance Assessment Platform (SEPAP)

The SEPAP is a cloud-based system created to establish a cohesive structure for evaluating and certifying energy performance through the integration of three evaluation approaches. It boasts intuitive interfaces, accommodates various languages, and incorporates a user support component [44]. Additionally, the platform employs a decision matrix that empowers engineers to automatically choose among the three evaluation methods based on specific building attributes like heating and cooling requirements, technical system intricacy, and on-site power production. SEPAP acts as a tool for merging artificial intelligence and machine learning methodologies to aid in adjusting building models to align with the capabilities of each approach. This process allows the platform to incorporate feedback from engineers' model modifications to improve its simulation of real-world scenarios [47]. Figure 3.2 shows the important features and capabilities of SEPAP.

The methods of assessment, which are divided into three parts known as modular approaches, can be defined as the subsequent strategies employed for evaluating and appraising various aspects within a given context [48]:

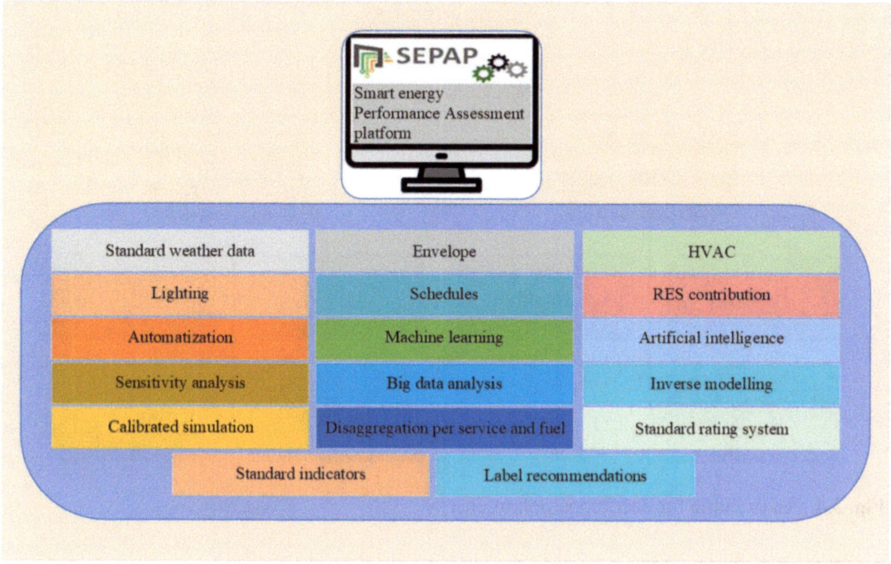

Fig. 3.2 Important features of SEPAP [48]

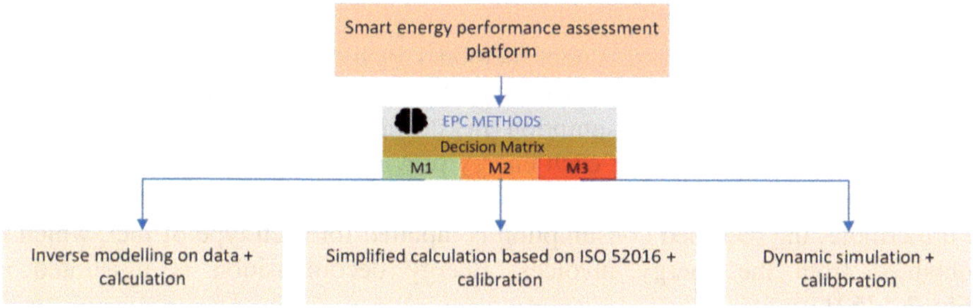

Fig. 3.3 The SEPAP assessment methods [51]

Method of Evaluation 1

Evaluation of energy performance using intelligent and data-informed methods.

Method of Evaluation 2

A simplified approach relies on the monthly computation (ISO 52016) interval and its adjustment [49].

Method of Evaluation 3

Sophisticated and automated simulation modeling relies on hourly computations (ISO 52017) and their adjustment [50]. Figure 3.3 shows the important tasks of each SEPAP method.

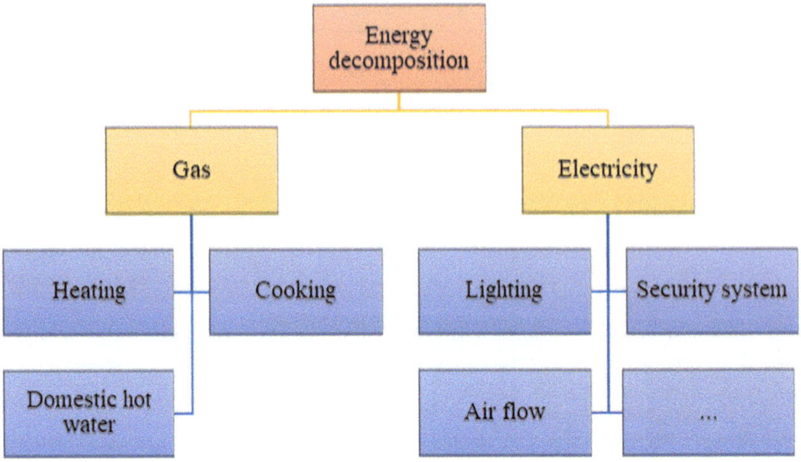

Fig. 3.4 An example for decomposition of energy

3.2.2.1 Method of Evaluation 1

The first assessment method involves smart energy management tools in buildings. This method employs inverse modeling to evaluate the energy performance of buildings.

The required data for assessment method 1 includes the energy use calculated through evaluation method 1, the necessary data for the SEPAP consists of the theoretically computed energy consumption (in kilowatt-hours [KWh]) for each service, and the availability of fuel for each energy provision. Additionally, the utilization data is entered for each fuel type, which is obtained from the energy invoices. Furthermore, the recorded consumption is inputted for each type of fuel, which is obtained from the energy invoices. Energy decomposition is illustrated in Fig. 3.4 [52].

3.2.2.2 Method of Evaluation 2

The subsequent evaluation approach within SEPAP, following ISO 52016, integrates intelligent technologies, artificial intelligence, and machine learning to conduct automated evaluations and sophisticated occupancy simulations. Fundamental building information such as geographical position, size, and structure category is provided. Meteorological information, timetables, building envelope specifics, HVAC installations, lighting, appliances, domestic hot water systems, and energy invoices are integrated. Users have the option to specify supplementary information.

The smart readiness indicator (SRI) is computed according to service performance tiers. The yearly energy usage is displayed in kilowatt-hours (kWh) [53].

3.2.2.3 Method of Evaluation 3

The third assessment approach incorporates advanced and automated simulation procedures that operate at an elevated level of complexity. The third evaluation technique, which is more intricate, involves sophisticated and automated simulation modeling that relies on hourly computations (ISO 52017) [50]. This method also entails calibration:

- The forecasting of building energy efficiency.
- Breakdown of energy usage by specific energy services and/or fuel types.
- Energy efficiency measures identification.
- Assessment of potential energy conservation and determination of economic feasibility.

Various technologies, such as the SEPAP tool built on Open Studio® [2], enable the automation of calibrating a dynamic simulation model for workflow processes. Users provide building data and energy usage specifics to compare with actual data, modifying behavior patterns to accurately measure energy savings [54].

3.3 New Energy Performance Certificate (EPC)

3.3.1 Concept of EPC

An energy performance certificate is mandatory for any non-domestic or commercial building upon construction, sale, or lease. This certificate provides information to potential occupants, tenants, or buyers regarding the energy efficiency of a building. The EPC allows individuals to include energy efficiency in their investment or business decision-making process. Also, EPC needs a recommendation report that outlines a series of activities to enhance the energy efficiency of the building. The suggestion report also provides an indication of the payback period for each of the actions [55]. The EPC remains valid for a maximum of 10 years. It is officially recorded and maintained in a national EPC database, allowing anyone interested in leasing or buying the property to access and assess its rating. Commercial properties available for sale, lease, or rent must also possess an EPC. The EPC may also be necessary for certain conversions and additions. The process for these is analogous for new constructions but commences with a site inspection to record the building specs and dimensions. The authorized software is additionally employed to compute the EPC rating for the structure. The evaluation of the certificate rating involves the analysis of the characteristics of the building envelope (fabric) and the building systems, encompassing ventilation, heating, and lighting [56].

3.3.2 The Performance of the EPC Method

The development of the new EPC template involved a participatory process that included workshops and interviews with multiple stakeholders. The EPC is seen by end users as a necessary document for contractual reasons, while other stakeholders see it as a beneficial policy tool for promoting energy transition in the building sector. End users in trial nations have a shared requirement for feedback on actual energy usage and accurate cost projections. This need is not met by the traditional EPC, which was originally intended to compare energy efficiency [57]. Experts recommend monitoring actual energy consumption; an additional tool is required to effectively address this matter. The EPC categorization consists of three levels: policy tool execution, EPC creation, and EPC application. An appropriate set of requirements for an effective EPC comprises reliability, comparability, understandability, tangibility, comprehensiveness, adequacy, and dynamism.

Here is some information that will make EPC more efficient:

Roadmap integration for renovation, energy costs for different content, relation between energy demand and CO_2 emissions, display of final energy demand and final energy consumption, and focus on the expenditure associated with the consumption of final energy from different energy sources. Figure 3.5 displays the energy efficiency rating (EER) and band (EER SAP, EERB SAP).

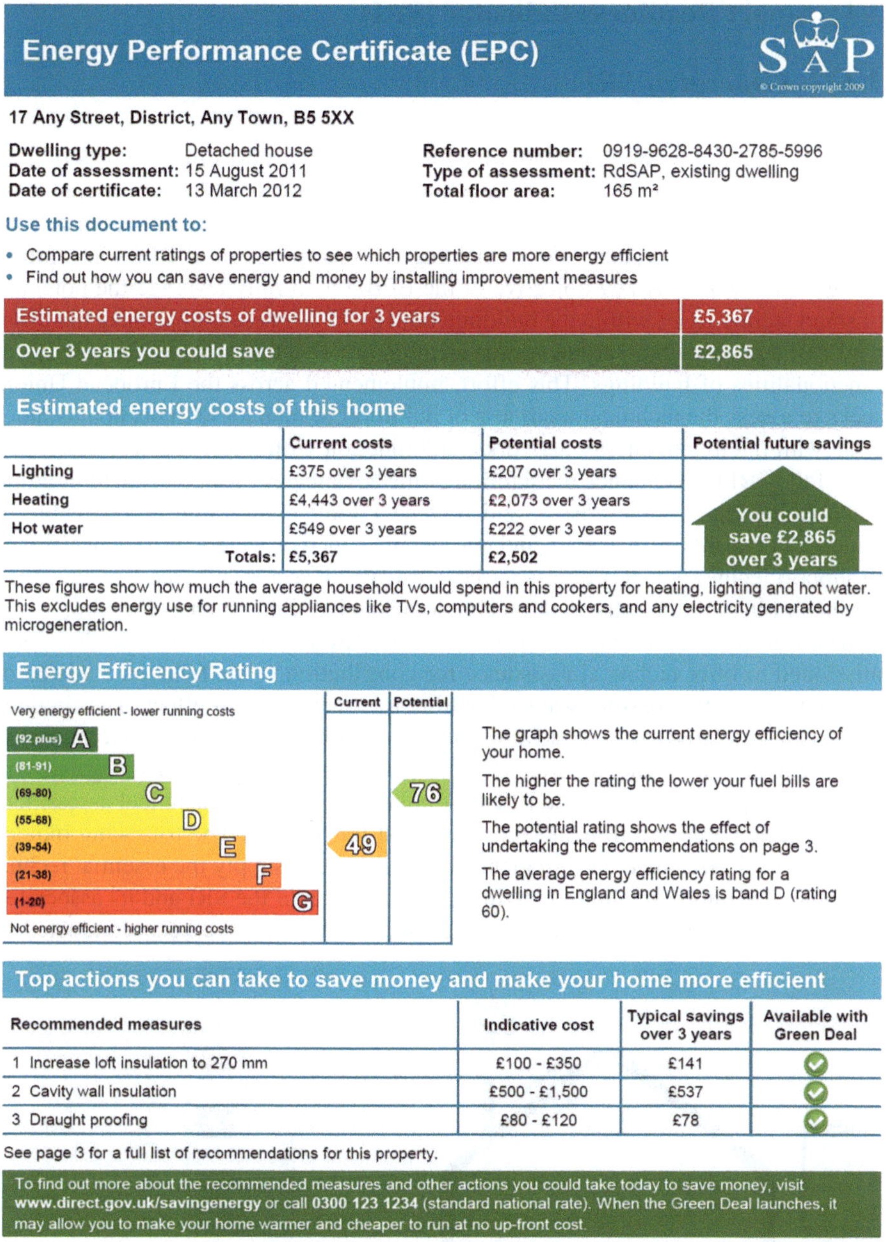

Energy Performance Certificate (EPC)

SAP © Crown copyright 2009

17 Any Street, District, Any Town, B5 5XX

Dwelling type: Detached house
Date of assessment: 15 August 2011
Date of certificate: 13 March 2012

Reference number: 0919-9628-8430-2785-5996
Type of assessment: RdSAP, existing dwelling
Total floor area: 165 m²

Use this document to:

- Compare current ratings of properties to see which properties are more energy efficient
- Find out how you can save energy and money by installing improvement measures

Estimated energy costs of dwelling for 3 years	**£5,367**
Over 3 years you could save	**£2,865**

Estimated energy costs of this home

	Current costs	Potential costs	Potential future savings
Lighting	£375 over 3 years	£207 over 3 years	You could save £2,865 over 3 years
Heating	£4,443 over 3 years	£2,073 over 3 years	
Hot water	£549 over 3 years	£222 over 3 years	
Totals:	£5,367	£2,502	

These figures show how much the average household would spend in this property for heating, lighting and hot water. This excludes energy use for running appliances like TVs, computers and cookers, and any electricity generated by microgeneration.

Energy Efficiency Rating

The graph shows the current energy efficiency of your home.

The higher the rating the lower your fuel bills are likely to be.

The potential rating shows the effect of undertaking the recommendations on page 3.

The average energy efficiency rating for a dwelling in England and Wales is band D (rating 60).

Top actions you can take to save money and make your home more efficient

Recommended measures	Indicative cost	Typical savings over 3 years	Available with Green Deal
1 Increase loft insulation to 270 mm	£100 - £350	£141	✓
2 Cavity wall insulation	£500 - £1,500	£537	✓
3 Draught proofing	£80 - £120	£78	✓

See page 3 for a full list of recommendations for this property.

To find out more about the recommended measures and other actions you could take today to save money, visit **www.direct.gov.uk/savingenergy** or call **0300 123 1234** (standard national rate). When the Green Deal launches, it may allow you to make your home warmer and cheaper to run at no up-front cost.

Fig. 3.5 Example of energy performance certificate (EPC) [58]

3.4 Smart Readiness Indicator (SRI)

3.4.1 Concept of SRI

Smart technologies implemented in buildings offer a cost-effective solution to enhance the creation of healthier and more comfortable building environments while reducing energy consumption and carbon footprint. Additionally, these technologies can play a crucial role in enabling the incorporation of RESs into future energy frameworks. A key objective of the EPBD is to fully leverage the potential of smart technologies within the building industry [59]. In order to achieve this goal, the EPBD includes instructions for the creation of a SRI that assesses the intelligent functionalities of buildings. This effort, implemented across the European Union, seeks to assess the technical readiness of buildings to interact with occupants, integrate with intelligent energy networks, and function with improved efficiency. The aim of the SRI is to enhance recognition of the advantages of sophisticated building technologies and characteristics and to render their extra value more visible to building occupants, proprietors, and intelligent service providers [60]. It also strives to support technological progress in the construction sector and create an incentive for integrating cutting-edge smart technologies in buildings such as Fig. 3.6.

The European Commission services (DG ENERGY) oversaw two studies commissioned to offer technical assistance for contributing to the talks on a standardized approach and possible ways of putting this measure into effect. The findings have been structured to aid the advancement of the SRI by the European Commission and Member States and to offer direction on formulating linked delegated and implementing actions as per the EPBD stipulations. An initial technical assessment has suggested a description and initial method for the SRI. The following inquiry into technical assistance has broadened existing data to supply the essential technical inputs for improving and finalizing the delineation of the SRI and its associated computation methodology. These two technical analyses have been conducted in

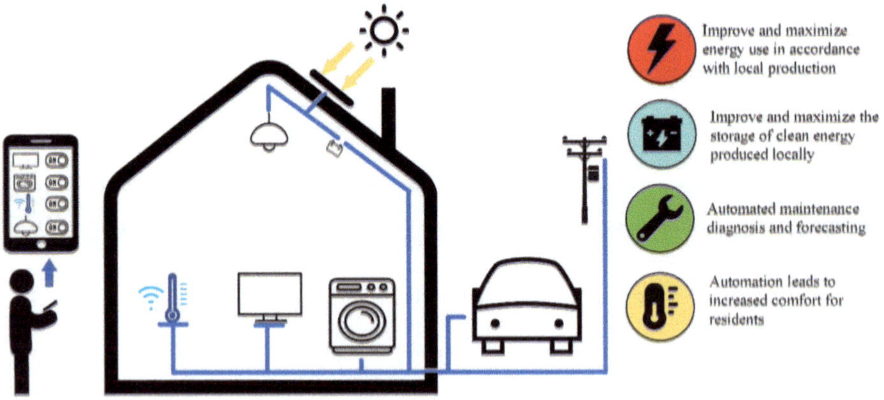

Fig. 3.6 Advantages of smart technologies in smart homes [61]

strong partnership with stakeholders, involving open discussions, five comprehensive stakeholder assemblies, surveys, a review of preliminary reports, and insights from three specialized stakeholder task forces.

The research group focusing on technical analysis discovered that the majority of individuals engaged concurred with the fundamental principles and methodological decisions of the SRI. A preliminary version of the approach underwent voluntary testing during a phase of open public evaluation, culminating in 112 evaluations carried out by engaged individuals throughout the EU. This activity affirmed the feasibility of the methodology and prompted further enhancements to the integrated approach. Additionally, the research delved into diverse strategies for executing the SRI, aiming to apprise the Commission Services and the Member States of potential frameworks for effectively executing the SRI program and its associated implications.

According to the EU's effect assessment, putting the SRI tool to use across the European Union could have big benefits for everyone [62]. Conclusively, the research on technical assistance has formulated and verified a feasible definition and evaluation technique for the SRI. The suggested method is in line with the goals specified in the EPBD, yields results that are sufficiently reliable, can be easily put into practice, and has proven to offer valuable insights to occupants of buildings. It has undergone thorough scrutiny and seems to garner widespread backing from various stakeholders, indicating its potential to serve as a suitable framework for facilitating the efficient execution of the SRI, potentially including additional assessments at the national level [63]. The evolution of SRI methods is shown in Fig. 3.7.

3.4.2 *Smart Readiness Indicator*

The phrase "smart readiness indicator" denotes an indicator that furnishes data regarding the degree to which a structure or a component thereof is prepared to operate smartly. The SRI for buildings was started based on the EPBD [25]. The goal of SRI is to provide a common process for evaluating buildings according to their technological preparedness for utilizing smart technology to improve energy efficiency, with comparable outcomes across EU member states [24]. The smart readiness indicator is designed to evaluate a building or building unit's ability to adjust its functioning to meet occupant and grid requirements, enhancing energy efficiency and overall performance. The content comprises enhanced energy efficiency, benchmarking criteria, flexibility, and cutting-edge functionalities derived from interconnected and smart gadgets. The smart readiness indicator will encompass the evaluation of a building or building unit's readiness for smart technologies, along with various readiness scores indicating the preparedness of buildings, building units, and systems across distinct essential functions, impact factors, and technical areas. [63]. The document will also strive to offer additional information regarding the inclusiveness and connectivity of the building, the interoperability and cybersecurity of systems, and data protection, whenever possible. The smart

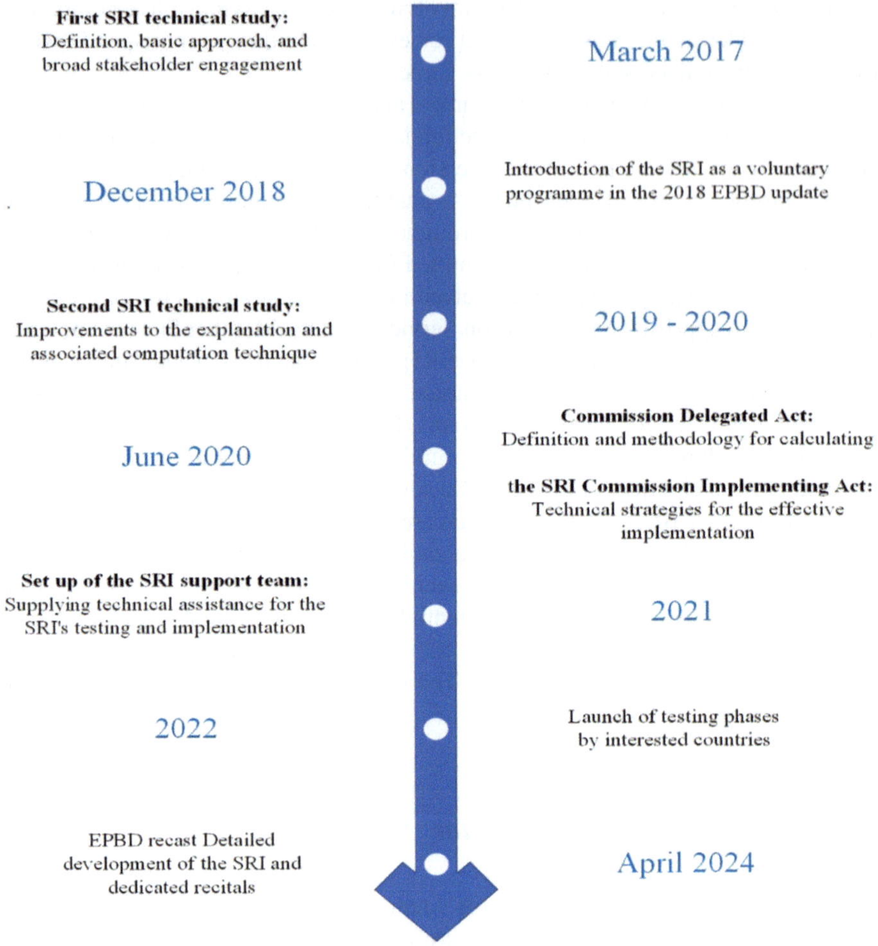

Fig. 3.7 History of the SRI [51]

readiness indicator should not serve as a measure of the energy efficiency of buildings. It is crucial for building owners to understand that the smart readiness indicator and energy performance certificates assess distinct aspects. Consequently, these aspects require unique approaches for improvement. Nonetheless, smart readiness has the potential to boost energy efficiency [62]. The calculation framework for the smart readiness indicator ought to be permissible for application across all categories of buildings specified in Directive 2010/31/EU [64]. The assessment of the intelligent readiness of constructions under the smart readiness indicator scheme for the provision of a smart readiness indicator certificate necessitates the engagement of proficient or endorsed professionals. Individuals endorsed in accordance with Directive 2010/31/EU for evaluating heating, ventilation, and air conditioning systems, or under Directive 2012/27/EU [65] of the European Parliament and Council for Energy audits, should be deemed competent to evaluate the smart readiness of constructions or building components as per Member States' discretion. With the proliferation of digitalization and connectivity within constructions, concerns regarding cybersecurity and data privacy are escalating. Consequently, constructions and their infrastructure are increasingly vulnerable to cyber threats and unauthorized access to personal data [66].

3.4.3 *The Optionality of the Plan*

Member states of the European Parliament have the autonomy to determine the extent of implementation of the smart readiness indicator plan within their territories. Residents in the EU retain the option to opt out of utilizing the plan. Additionally, states have the flexibility to selectively apply the smart readiness indicator plan to specific buildings. The decision to mandate compliance for certain buildings or building units within their jurisdiction rests with the states, allowing individuals the choice to abstain. Prior to implementing the plan across all or a portion of their territories, states are required to inform the Commission accordingly. States may also choose to alter, customize, or stop the program's implementation at any time without giving a reason [13]. Figure 3.8 shows an example of a SRI certificate. In conclusion, the preferred certificate layout would encompass the following elements:

- the comprehensive SRI rating
- the SRI category (derived from the structure's total SRI rating using a seven-level system like A to G)
- the SRI percentages for the three primary functions—the SRI percentages for all seven impact criteria
- the grid showing SRI ratings categorized by technical field and impact criteria

In a building's technical domain, full utilization of smart features results in a 100% rating for the impact criterion, while no smart features yield a 0% rating. Varying levels of smartness correspond to values between 0 and 100%. If all technical domains in a building achieve 100% ratings for a specific impact criterion, the building's overall score for that criterion would also be 100% [68].

3.4.4 *Benefits of SRI*

There is an essential need to accelerate investments in the refurbishment of structures in Europe by integrating smart, energy-efficient technologies. Smart buildings leverage advanced information and communication technology solutions to efficiently regulate energy consumption, improve comfort, and reduce energy usage. The expected surge in the adoption of smart technologies is forecasted to deliver economical energy savings and heightened indoor convenience. The importance of

Fig. 3.8 An example of SRI certificate [67]

smart buildings is crucial for future energy frameworks that will heavily depend on renewable sources. The EPBD underscores Smart Ready Technologies by introducing the "SRI" to evaluate the level of sophistication in buildings for occupants and service providers.

The purpose of the SRI scheme is to increase knowledge regarding the advantages of smart buildings, namely in terms of energy efficiency. This will encourage investments in smart building technologies and facilitate their adoption. According to technological advancements in the construction industry, the SRI prefers to find better ways for policymakers in energy, buildings, and other areas to work in common, especially when it comes to ICT (information and communication technology). This will make it easier for the building industry to join energy markets and processes in the future [27, 69].

3.4.5 SRI Service Catalog

The SRI approach evaluates smart-ready services in a technology-agnostic manner. This evaluation utilizes a checklist-style inventory that can be employed through self-assessment, external audits, or data gathering on equipment. A functionality level ranging from 0 to 4 is chosen for each assessment aspect (service) to depict the actual functionality accurately [70].

3.5 Overview of the Scoring Methodology

When assessing structures or components within the SRI framework, the focus is on evaluating their compliance with three key criteria and capabilities. Figure 3.9 shows seven impact criteria that can be used to further elaborate on the three basic smart readiness features [71].

3.5.1 Energy Efficiency

This specific impact category pertains to the effects of smart-ready services on the capacities of energy conservation. Only the impact of smart-ready technologies, such as improved control of room temperature settings, is considered when evaluating the overall energy performance of buildings [72].

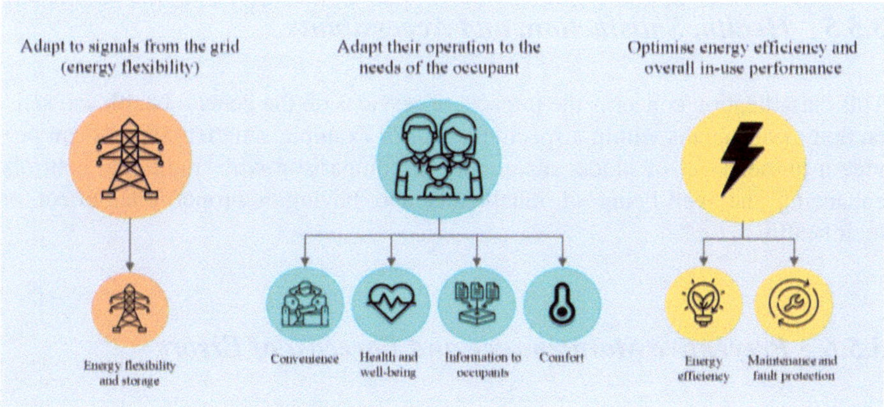

Fig. 3.9 Key functionalities and seven impact criteria [51]

3.5.2 Energy Flexibility

This domain pertains to the impact of services on the building's capacity for modify its energy utilization. The proposed study aims to expand its scope beyond electricity grids by incorporating flexibility into district heating and cooling grids as well [63].

3.5.3 Comfort

This group of impacts focuses on the way services make users feel more comfortable. The notion of comfort encompasses both deliberate and automatic reactions to the environment. It covers aspects like thermal comfort, auditory comfort, and visual functionality, which pertain to obtaining adequate illumination devoid of harsh reflections.

3.5.4 Simple Access

This domain of influence pertains to the effects of services on the convenience of individuals, specifically the degree to which services streamline their everyday tasks, like Technical Building Systems (TBS) reducing the need for manual engagements [60].

3.5.5 Health, Satisfaction, and Accessibility

This classification concerns the impacts of services on the general health and satisfaction of occupants within a specific area. For example, smarter controls can provide a higher level of indoor air quality in comparison with traditional controls, enhancing the well-being of inhabitants and having a proportional effect on their health.

3.5.6 Preventive Maintenance and Forecast of Errors

It is possible that using automated fault detection and diagnosis procedures will make technical building systems much easier to operate and maintain more efficiently. It can also change how energy-efficient technical building systems are by finding and fixing processes that aren't working as well as they could.

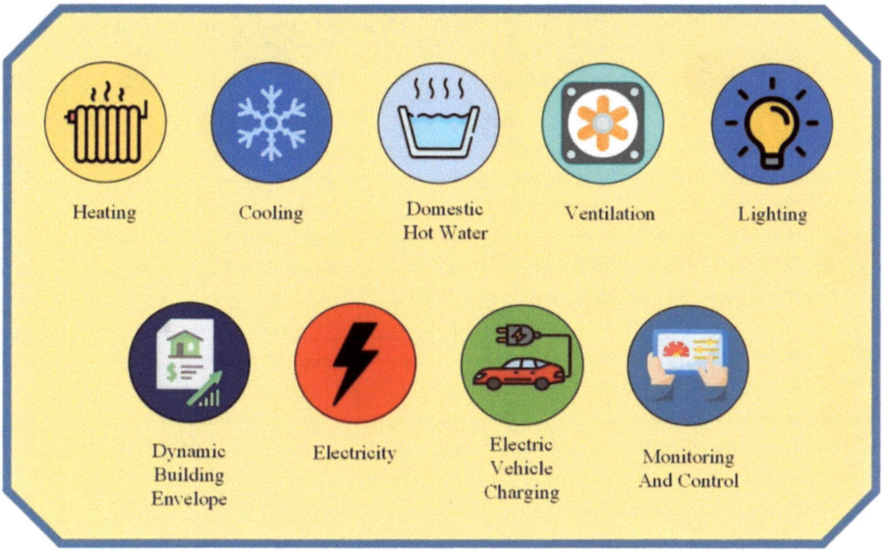

Fig. 3.10 Nine technical domains [51]

3.5.7 Details for the Residents

This impact category is concerned with how services affect inhabitants' access to information about building operations [27].

The SRI calculation approach relies on evaluating the smart-ready services that the building now possesses or has the potential to utilize, referred to as the "service catalog." These services are divided into nine technical domains, as shown in Fig. 3.10.

The assessment yields comprehensive results (up to 57 ratings) broken down by effect criterion and domain. Further, the results for each of the nine technological domains are combined. Total points for every one of the seven impact criteria and total points for all three main features are shown in Fig. 3.11. As well as the total SRI score and the matching SRI class (there are seven classes, ranging from SRI 0% to 100%). Figure 3.11 illustrates how to evaluate the different levels of seven impact criteria and nine technical domains [73].

There are two methods to assess SRI, which are Method A (simplified) and Method B (detailed), which are shown in Fig. 3.12.

The catalogs of SRI service are structured to classify services across nine specific areas: heating, cooling, hot water supply, controlled ventilation, lighting, dynamic building envelope, power supply, electric vehicle charging, and monitoring and regulation. The member state conducting the SRI test or implementation phase will decide whether to use regular methods or customized methods (A or B). This decision will be based on the specific features of that country. Two to five functionality levels are defined for each of the services. A higher level of functionality

	Overall SRI score(%) + SRI class						
	Optimize energy efficiency and overall in use performance		Adapt its operation to the need of the occupant				Adapt to signals from the grid (energy flexibility)
	Energy efficiency	Maintenance and fault protection	Comfort	Convenience	Health and well-being	Information to occupants	Energy flexibility and storage
Total	39 %	18%	60%	71%	48%	59%	51%
HEATING	32%	18%	62%	55%	24%	74%	100%
COOLING	65%	51%	78%	72%	61%	55%	0%
DOMESTIC HOW WATER	17%	0%	45%	70%	67%	83%	0%
VENTILATION	41%	0%	55%	60%	34%	44%	--
LIGHTING	85%	14%	90%	100%	83%	15%	--
DYNAMIC BUILDING ENVELOPE	10%	0%	31%	56%	22%	46%	--
ELECTRICITY	10%	0%	--	--	--	68%	0%
ELECTRIC VEHICLE CHARGING	--	38%	--	82%	--	84%	25%
MONITORING AND CONTROL	52%	43%	62%	72%	45%	64%	14%

Fig. 3.11 Example of scores calculated at different levels [51]

Method A (simplified)

- Service Catalog Simplified
- Usually, there are residental buildings and small non-residental buildings with low complexity
- Approach based on a checklist
- Time of assessment < 1 hour
- Self-assessment is achievable
- The differentiation within the service catalog requires a corresponding degree of proficiency for conducting the evaluation
- These two methods use the same assessment method

Method B (detailed)

- Catalog of full-detail service
- Usually for new residential buildings and non-residential buildings with high complexity
- On-site assessment is required
- Time of assessment < 1 day
- Requires an expert or the presence of a facility
- Differentiating items in the service catalog necessitates a proportional level of expertise to carry out the assessment.
- These two methods use the same assessment method

Fig. 3.12 Two methods of calculating SRI [61]

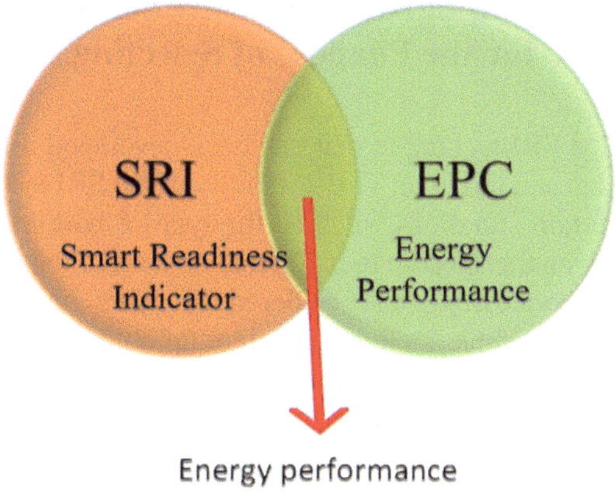

Fig. 3.13 SRI and EPC complement each other [51]

indicates a more smart implementation of the service, typically yielding more advantages for building users or the grid in comparison to services implemented at a lower level of functionality. The degrees of functionality are represented as ordinal numbers, indicating that it is not possible to directly compare ranks between different services in a quantitative manner. The individuals responsible for the issuance of EPCs possess the necessary expertise to also issue SRI certifications. Also, the issuance of the SRI certificate may be coupled with the EPC scheme (which is shown in Fig. 3.13) by the member states [24].

The smart readiness score of a building, as per the proposed SRI methodology, is a percentage that indicates the proximity or distance of the building to achieve maximum smart readiness. A larger proportion indicates a higher level of smartness in the building. The conversion of the % to an alternative indicator, such as a star rating or alphabetical score (A, B, C, etc.), is also possible. This hypothesis has been subjected to more scrutiny with graphical representations and the conduct of market research involving specific customer groups. The initial step involves evaluating each smart-ready service. The services provided within the building undergo inspection to assess their level of operation. An effect score is assigned to each service based on the impact parameters examined in the approach.

After determining the effect scores of each individual service, a combined impact score is computed for each domain examined in the methodology. The domain-impact score is determined by dividing the individual scores of the domains' services by the theoretical maximum individual scores, presented as a percentage. The total impact score is determined by calculating a weighted sum of the domain-impact scores for each impact criterion. The weight assigned to a certain domain in this calculation will be contingent upon its relative significance in relation to the impact under consideration [69].

3.6 Standardized the SRI Score and Triage Process for the Purpose of Selecting Appropriate Services

The SRI technique that has been developed offers a framework that is both adaptable and modular. The suitability of the SRI technique is expected to differ based on certain situations, such as the type of building, climate, and site-specific factors. Given the specific location and context of the site, some areas, offerings, and levels of service might be considered redundant, inappropriate, or unacceptable. Therefore, the SRI must include sufficient flexibility to meet these circumstances. The maximum nominal impact score is not solely determined by adding together the effects of the services offered in the simplified SRI catalog. It is probable that certain domains and services may lack relevance, applicability, or desirability due to the specific local and site-specific context. The SRI method effectively addresses this issue through the utilization of a triage process to determine the appropriate services for a specific establishment [74]. Some standards are shown in Fig. 3.14.

Certain areas might not have significance, like specific structures without the ability to provide parking spaces, thus restricting their potential to supply electric vehicle charging. Some residential buildings may not require cooling systems at all. Moreover, numerous services rely on particular technical building infrastructures, such as a hot water storage facility or a heat recovery ventilation system. It's also important to preserve in that some services may not work via individuals. For example, a building probably can't have both district heating and combustive heating and heat systems at the same time. If these services are not there, it's clear that they don't need to be looked at during on-site checks. There are a lot of factors that affect how many services need to be checked during an SRI review in a real building. The SRI catalog contains fewer than 54 smart-ready services (or 27 under Method A).

The triage procedure has an impact not only on the duration and exertion of inspections but also on the attainment of the highest possible score. It would be unjust to impose penalties on a building for failing to provide services that are not pertinent. The SRI should refrain from encouraging intricacy in buildings and thus will solely consider services that are either existing or preferable. For certain services, this can vary depending on the individual scenario. In this scenario, a passive dwelling equipped with sunshades, ventilation, and/or window control mechanisms may eliminate the necessity for artificial cooling and should not be excluded based on the absence of such features.

Essentially, two methods for dealing with missing domains or services are merged [61]:

- Certain services only need to be assessed when the necessary technical building systems are in place (thus making them "smart ready"). In situations where assessors are unable to definitively identify the domain's relevance, this approach is appropriate. For example, the architecture of the building (direction, window-to-wall ratio, etc.) has a significant impact on the usefulness of automated shading systems. The SRI does not allow for the objective making of such an assessment. However, the SRI can assess how smartly the shading devices control when the shades are changed.

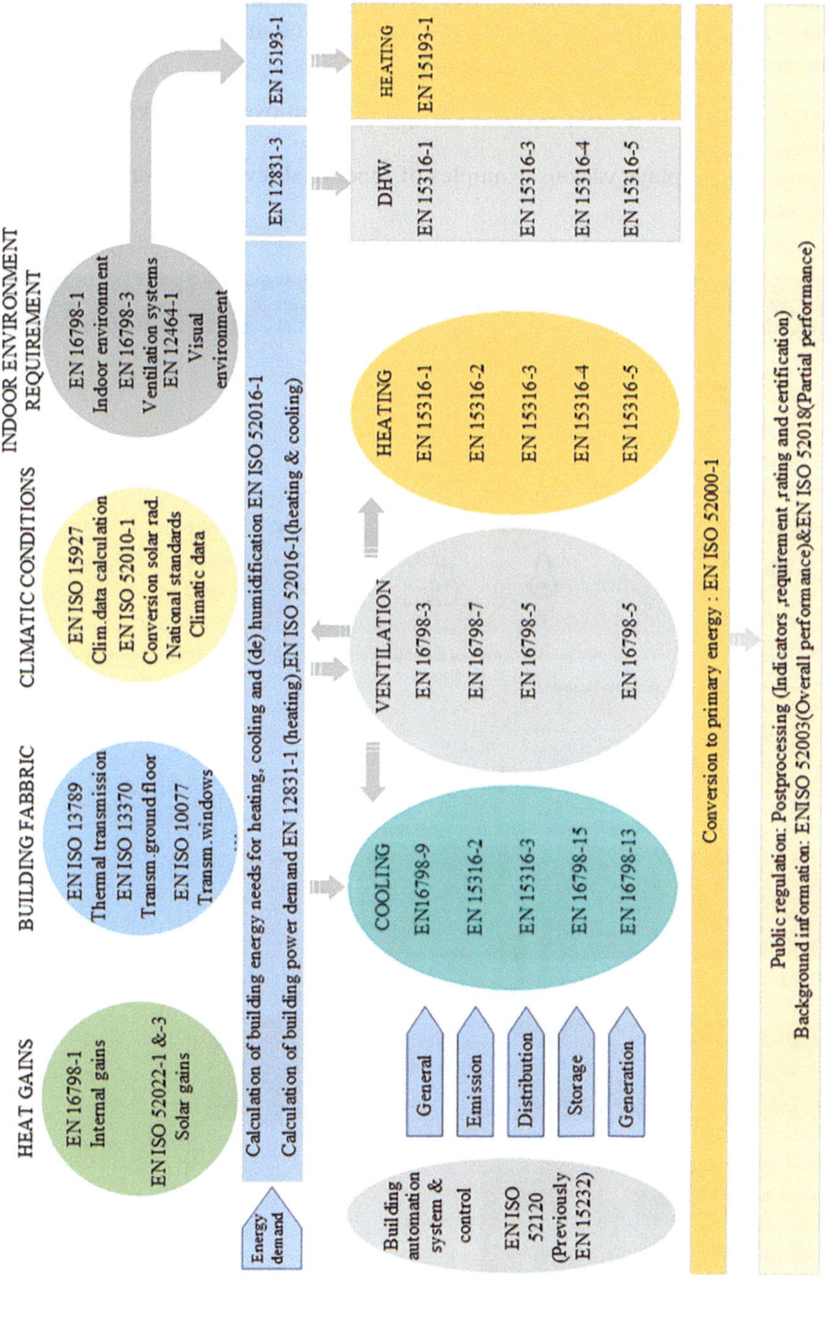

Fig. 3.14 The standard's placement inside the CEN and ISO energy performance of buildings (EPB) standard set [75]

- Some services might not exist, but they are nevertheless desirable from a policy standpoint, which is why they are "smart possible." This strategy could act as a catalyst for adding more smart services to existing buildings. To increase the SRI score, for example, punishing the lack of a controlled ventilation system may encourage the installation of one.

Figure 3.15 shows a brief explanation and gives an overall overview of how to calculate.

Figure 3.16 displays various examples of functional levels accompanied by different codes.

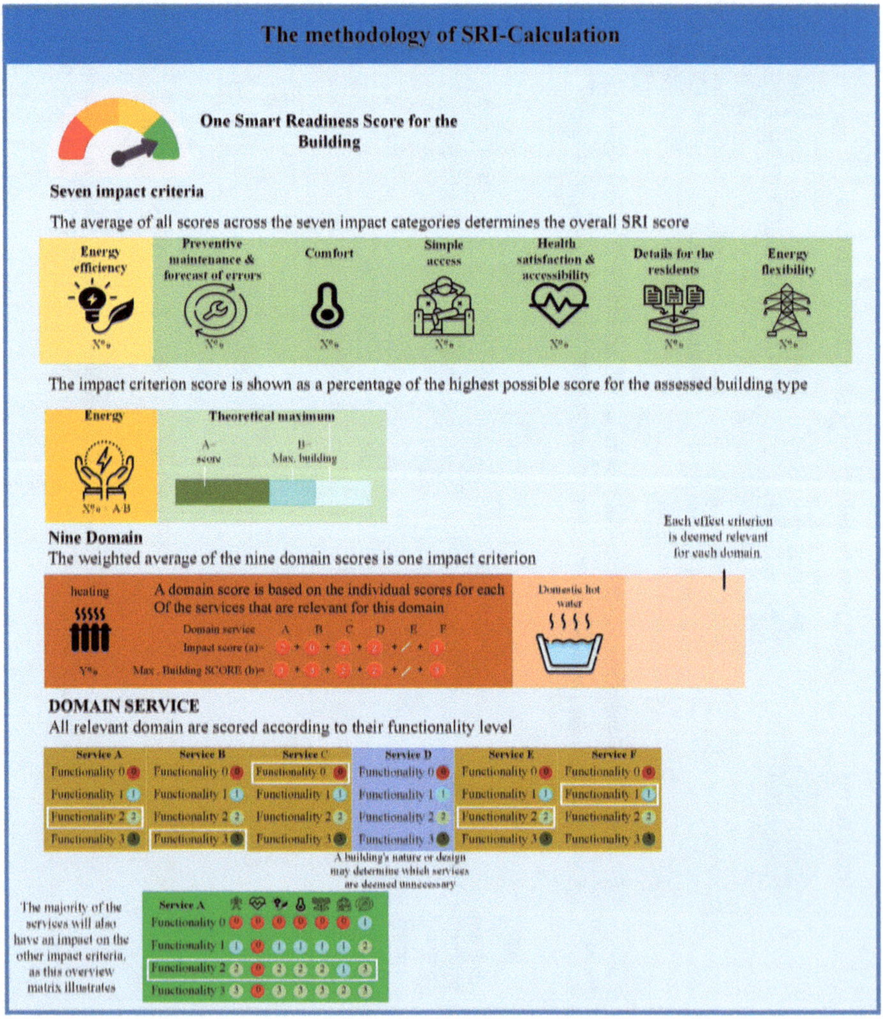

Fig. 3.15 Brief explanation of the way to calculate with example [61]

Code	Smart ready service	level 0 (Non Smart)	Level 1	Level 2	Level 3	Level 4 (Maximum Smart)
H-1a	Heat emission control	Without automatic control	Central automatic control	Individual room control	Each room has its own control system that can interact with other controllers and the Building Automation and Control System (BACS)	Each room has the ability to be controlled individually, with the capability to communicate and detect occupancy.
DHW-3	Information of regarding domestic hot water performance	-------	Indication of actual values	Actual values and historical data	Performance evaluation including forecasting and/or benchmarking	Performance evaluation including forecasting and/or benchmarking; also including predictive management and fault detection
C-2a	Generator control for cooling	On/Off-control of cooling production	Multi-stage control of cooling production. capacity depending on. the load or demand	Adjusting the cooling production capacity based on the load or demand through variable control	Variable control of cooling production capacity depending on the load and external signals from grid	------
V-1a	Supply air flow control at the room level	No ventilation system or manual control	Clock control	Occupancy detection control	Central Demand Control based on air quality sensors	Local Demand Control based on air quality sensors (CO2, VOC,...) with local flow from/to the zone regulated by dampers
L-1a	Occupancy control for indoor lighting	Manual on/off switch	Manual on/off switch + additional sweeping extinction signal	Automatic detection	Automatic detection	-------

Fig. 3.16 Example of each stage with different codes [51]

DE-1	Window solar shading control	No sun shading or only manual operation	Motorized operation with manual control	Motorized operation with automatic control based on sensor data	Combined light-blind-HVAC control	Predictive blind control (e.g. based on weather forecast)
E-2	Reporting information regarding local electricity generation	--------	Available data from the current generation	Existing values and historical information	Performance review, which includes forecasting and/or benchmarking	Evaluations of performance include forecasting, benchmarking, predictive management, and fault detection
EV-15	Capacity of EV charging	---------	Ducting (or simple power plug) available	Under 9% of parking spaces has recharging points	10 to 50% or parking spaces has recharging point	More than 50% of parking spaces has recharging point
MC-25	Integration of Smart Grid	None-No harmonization between grid and TBS; The building operates self-sufficiently, separate from the grid's electrical consumption	Demand side management (DSM) possible for (some) individual TBS, but not coordinated over various domains	Coordinated demand side management (DSM) of multiple TBS	-------	---------

Fig. 3.16 (continued)

3.7 SRI Score Calculation

The cumulative intelligent readiness rating is established through a systematic method. The ultimate SRI value is presented as 0–100% and illustrates the proportional intelligence in relation to a completely smart structure. Evaluation and computation of the SRI score are carried out following a series of steps [76, 77]:

- The initial phase involves pinpointing and incorporating pertinent services within the designated structure while disregarding irrelevant services to prevent the penalization of buildings due to the absence of irrelevant domains and services. In this case, a building that uses DH for heating is unlikely to have additional combustion or heat pump heating systems. Therefore, services related to these systems are not included. This process is called the triage method.

- Relevant services undergo evaluation by choosing from functionality tiers ranging from 0 to 4, with level 0 representing the foundational tier and level 4 denoting the highest degree of advancement. Depending on the tiers chosen, services impact one or multiple SRI criteria.

Service point distribution tables can be found within the SRI catalog. For instance, the allocation of points for the service labeled "Ventilation-1a" adheres to the guidelines specified in Fig. 3.17. These allocations are primarily guided by established norms, particularly concerning the evaluation aspect of "energy efficiency." In contrast, when considering criteria like "convenience" or "information provided to occupants," there is a scarcity of standardized benchmarks, leading to distributions that are somewhat influenced by subjective assessments. The relationship between the impact criteria and the functionality levels of service "Ventilation-1a" is made clear in Fig. 3.17 that certain impact factors, including "energy flexibility and storage," have 0 points assigned to all levels; hence, the service has no bearing on these criteria. The maximum impact points that can be obtained for other impact criteria, including "energy efficiency," are higher than zero (three points at level 4). As a result, the impact criterion scores are impacted by the functionality level that is chosen. In the following stage, the aggregation process is started when all services for a domain have been evaluated.

- Equation (3.1) is employed to compute the impact criteria point score for every domain-impact combination. This process entails the accumulation of relevant

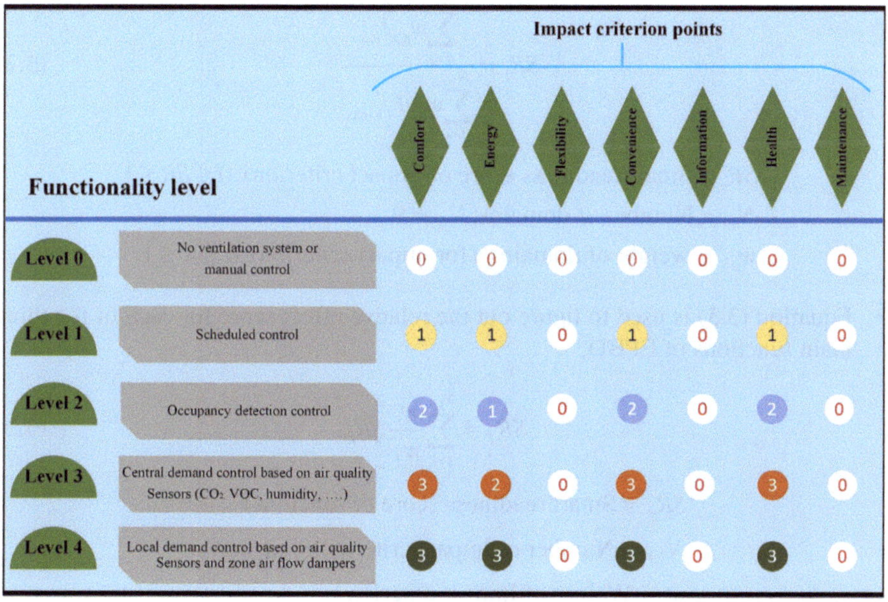

Fig. 3.17 Service "Ventilation-1a supply airflow control at the room level" gave the award for impact factor $I_{(s,d,i)}$. Each service has its own set of points that can be used [76]

service points, with the exclusion of any services that were initially omitted. The computation of points for seven impact criteria across nine domains results in a cumulative score of 63 points. Lowercase subscripts are employed to represent numerical indices of specific services, domains, impact criteria, and functions in the equations above. On the other hand, subscripts in capital letters are utilized to symbolize categories.

$$I_{d,i} = \sum_{S=1}^{N_{s,d}} I_{s,d,i} \tag{3.1}$$

$I_{d,i}$ = Point score of impact criterion i in domain $d; I \in \mathbb{Z}$

$N_{s,d}$ = Number of service in domain $d; N \in \mathbb{N}$

$I_{s,d,i}$ = Point score of service s in domain d for impact criterion i

s,d,i = Numerical index of service, domain or impact; $(s,d,i) \in \mathbb{N}$

- Maximum score for the impact factor for each domain, the impact combination $I_{d,\,i,\,max}$ is found using Eq. (3.1) again, but again, the achieved service point scores are used instead. $I_{s,\,d,\,i}$ with the highest scores $I_{d,\,i,\,max}$. One can get the numbers from, assume, the highest level 4 values in Fig. 3.17.
- Equation (3.2) is used to figure out the smart readiness impact score for each of the seven impact factors. Also, in the following numerical values of weighting factors are explained. Relative values are shown as numbers between 0 and 1.

$$SR_i = \frac{\sum_{d=1}^{N_D} w_{d,i} I_{d,i}}{\sum_{d=1}^{N_D} w_{d,i} I_{d,i,\max}} \tag{3.2}$$

SR_i = Smart readiness score of impact criterion i; $0 \le SR \le 1$

N_D = Number of domains; $N_D = 9$

$w_{d,i}$ = weight of domains d for impact criterion $i; 0 \le w \le 1$

- Equation (3.3) is used to figure out the relative effect score for each of the three main functions of EPBD.

$$SR_f = \sum_{i=1}^{N_{1,f}} \frac{w_i}{w_f} SR_i \tag{3.3}$$

SR_f = Smart readiness score of function f

$N_{1,f}$ = Number of impact criteria in function f

w_i = Weight of impact criterion i

w_f = Weight of EPBD function f

- Equation (3.4) illustrates how the overall smart readiness score is computed by averaging the function or impact criterion scores with respect to weight.

$$SR = \sum_{i=1}^{N_1} w_i SR_i \tag{3.4}$$

SR = The total smart readiness score

N_1 = Number of impact criteria; $N_1 = 7$

- Equation (3.5) can be used to calculate relatively smart readiness scores for each of the 63 domain-impact pairs, if needed.

$$SR_{d,i} = \frac{I_{d,i}}{I_{d,i,\max}} \tag{3.5}$$

$SR_{d,i}$ = Smart readiness score of selected domain impact pair

3.8 Weighting of Technical Domains

The impact criteria allocate weights to each technical domain, with the weighting variables determining the extent to which the technical domain affects the criterion.

- The percentage values representing the weighting factors allocated to various technical domains are displayed. The total of the weighting factors pertaining to the technical domains for every impact criterion equals 100%.
- Weighting factors for technical domains are traditionally assigned using the following method:
  - The climatic zone's energy equilibrium considers aspects like "warming," "cooling," "hot water for household use," "air circulation," "illumination," and "electricity" within the technical domains of "energy effectiveness," "maintenance and prediction," and "energy flexibility and retention."
  - Weighting factors can be either constant or uniformly distributed in all other cases.
- If needed, member states decide which climate zones are used to figure out the weighting factors. Member States are allowed to use relevant Union guidelines if they are available.
- Weighting factors for technical domains pertaining to specific impact criteria may vary between residential and non-residential buildings.
- The Union offers relevant suggestions for Member States to consider when determining the weighting coefficients, along with contemplating the potential impacts of climate variability [64].

3.9 Conclusion

This study emphasizes the pivotal role of the smart readiness indicator in enhancing the energy performance of buildings within the framework of the EPBD. The SRI serves as a comprehensive tool for assessing the integration of smart technologies in buildings, thereby facilitating a transition towards more sustainable urban environments. The findings indicate that the implementation of the SRI is expected to yield significant benefits, including improved energy efficiency and enhanced occupant comfort. The central emphasis of this idea lies in the focus on monitoring and control, guaranteeing the seamless integration and effective management of all building systems.

Moreover, the research underscores the necessity of a standardized approach to the SRI, which has garnered widespread support from various stakeholders through extensive consultations and evaluations. The proposed methodology not only aligns with the overarching goals of the EPBD but also provides reliable and actionable insights for building occupants and policymakers alike. As the European Union moves towards its decarbonization targets for 2050, the SRI stands out as a critical instrument for driving innovation and fostering a culture of sustainability in the built environment. The chapter advocates for the continued refinement and implementation of the SRI, highlighting its potential to transform building performance assessments and contribute to a greener future.

References

1. B. Ramezani, M.G.D. Silva, N. Simões, Application of smart readiness indicator for Mediterranean buildings in retrofitting actions. Energ. Buildings **249**, 111173 (2021)
2. A. Safari, M.M. Hayati, M. Nazari-Heris, Hydrogen-Combined Smart Electrical Power Systems: An Overview of United States Projects, in *Green Hydrogen in Power Systems*, ed. by V. Vahidinasab, B. Mohammadi-Ivatloo, J. Shiun Lim, (Springer International Publishing, Cham, 2024), pp. 321–340
3. International Energy Agency (IEA), World Energy Outlook 2023, CC BY 4.0 (report); CC BY NC SA 4.0 (Annex A), October 2023. https://www.iea.org/reports/world-energy-outlook-2023. Accessed 25 April 2024.
4. D. Connolly, H. Lund, B.V. Mathiesen, Smart energy Europe: The technical and economic impact of one potential 100% renewable energy scenario for the European Union. Renew. Sust. Energ. Rev. **60**, 1634–1653 (2016)
5. M.M. Hayati, H. Majidi-Gharehnaz, H. Biabani, A. Aminlou, M. Abapour, An Overview of Energy and Exergy Analysis for Green Hydrogen Power Systems, in *Green Hydrogen in Power Systems*, ed. by V. Vahidinasab, B. Mohammadi-Ivatloo, J. Shiun Lim, (Springer International Publishing, Cham, 2024), pp. 1–24
6. M. Daneshvar, M. Pesaran, B. Mohammadi-ivatloo, Transactive Energy in Future Smart Homes, in *The Energy Internet*, (Elsevier, Amsterdam, 2019), pp. 153–179
7. Policy innovation for building renovation. How can policy innovation scale up the decarbonisation of the building stock in Europe, Buildings Performance Institute Europe (BPIE) December 2018. https://bpie.eu/wp-content/uploads/2019/01/BPIX-Briefing-_Final-1.pdf. Accessed 27 April 2024.

8. J. Al Dakheel, C. Del Pero, N. Aste, F. Leonforte, Smart buildings features and key performance indicators: A review. Sustain. Cities Soc. **61**, 102328 (2020)
9. X. Cao, X. Dai, J. Liu, Building energy-consumption status worldwide and the state-of-the-art technologies for zero-energy buildings during the past decade. Energ. Buildings **128**, 198–213 (2016)
10. Y. Noh, S. Jafarinejad, P. Anand, A review on harnessing renewable energy synergies for achieving urban net-zero energy buildings: Technologies, performance evaluation, policies, challenges, and future direction. Sustainability **16**(8), 3444 (2024)
11. C. Maduta, D. D'Agostino, S. Tsemekidi-Tzeiranaki, L. Castellazzi, G. Melica, P. Bertoldi, Towards climate neutrality within the European Union: Assessment of the energy performance of buildings directive implementation in member states. Energ. Buildings **301**, 113716 (2023)
12. M. De Groote, J. Volt, F. Bean, Is Europe ready for the smart buildings revolution, 2017
13. M.C. Pinto, Towards smart buildings. Is the smart readiness indicator an effective tool, Politecnico di Torino, 2020
14. S. Arun, M. Selvan, Intelligent residential energy management system for dynamic demand response in smart buildings. IEEE Syst. J. **12**(2), 1329–1340 (2017)
15. Y. Zhou, S. Cao, J.L. Hensen, P.D. Lund, Energy integration and interaction between buildings and vehicles: A state-of-the-art review. Renew. Sust. Energ. Rev. **114**, 109337 (2019)
16. R. Selvaraj, V.M. Kuthadi, S. Baskar, Smart building energy management and monitoring system based on artificial intelligence in smart city. Sustain Energy Technol Assess **56**, 103090 (2023)
17. E. Janhunen, N. Leskinen, S. Junnila, The economic viability of a progressive smart building system with power storage. Sustainability **12**(15), 5998 (2020)
18. V. Apostolopoulos, P. Giourka, G. Martinopoulos, K. Angelakoglou, K. Kourtzanidis, N. Nikolopoulos, Smart readiness indicator evaluation and cost estimation of smart retrofitting scenarios - A comparative case-study in European residential buildings. Sustain. Cities Soc. **82**, 103921 (2022)
19. R. Rameshwar, A. Solanki, A. Nayyar, B. Mahapatra, Green and Smart Buildings: A Key to Sustainable Global Solutions, in *Green Building Management and Smart Automation*, (IGI Global, New York, 2020), pp. 146–163
20. Buildings The global exchange for energy efficiency policies, data and analysis, 2019. https://www.iea.org/topics/energyefficiency/buildings/. Accessed 27 April 2024
21. M.M. Hayati, B. Motallebi Azar, A. Aminlou, M. Abapour, K. Zare, Techno-Economic Analysis for Centralized GH2 Power Systems, in *Green Hydrogen in Power Systems*, ed. by V. Vahidinasab, B. Mohammadi-Ivatloo, J. Shiun Lim, (Springer International Publishing, Cham, 2024), pp. 59–83
22. M.H. Rehmani, M. Reisslein, A. Rachedi, M. Erol-Kantarci, M. Radenkovic, Integrating renewable energy resources into the smart grid: Recent developments in information and communication technologies. IEEE Trans. Industr. Inform. **14**(7), 2814–2825 (2018)
23. E. O'Dwyer, I. Pan, S. Acha, N. Shah, Smart energy systems for sustainable smart cities: Current developments, trends and future directions. Appl. Energy **237**, 581–597 (2019)
24. P.A. Fokaides, C. Panteli, A. Panayidou, How are the smart readiness indicators expected to affect the energy performance of buildings: First evidence and perspectives. Sustainability **12**(22), 9496 (2020)
25. I. Vigna, R. Pernetti, G. Pernigotto, A. Gasparella, Analysis of the building smart readiness indicator calculation: A comparative case-study with two panels of experts. Energies **13**(11), 2796 (2020)
26. Directive (EU) 2018/844 of the European Parliament and of the Council of 30 May 2018 amending Directive 2010/31/EU on the energy performance of buildings and Directive 2012/27/EU on energy efficiency (Text with EEA relevance). https://eur-lex.europa.eu/eli/dir/2018/844/oj. Accessed 23 April 2024
27. K. Chatzikonstantinidis, E. Giama, P.A. Fokaides, A.M. Papadopoulos, Smart readiness indicator (SRI) as a decision-making tool for low carbon buildings. Energies **17**(6), 1406 (2024)

28. Consolidated text: Directive 2010/31/EU of the European Parliament and of the Council of 19 May 2010 on the energy performance of buildings (recast), 2010
29. T. W. E. relevance, Directive (EU) 2023/1791 of the European Parliament and of the Council. Official Journal of the European Union, 13 September 2023
30. E. relevance, Directive (EU) 2024/1275 of the European Parliament and of the Council, of 24 April 2024
31. C.A. Bollino, F. Botti, Energy poverty in Europe: A multidimensional approach. PSL Quart. Rev. **70**(283), 473–507 (2017)
32. G. Battista, L. Evangelisti, C. Guattari, C. Basilicata, R. de Lieto Vollaro, Buildings energy efficiency: Interventions analysis under a smart cities approach. Sustainability **6**(8), 4694–4705 (2014)
33. Directive 2010/31/EU of the European Parliament and of the Council of 19 May 2010 on the energy performance of buildings," Official Journal of the European Union, 19 May 2010
34. Directive (EU) 2018/844 of the European Parliament and of the Council of 30 May 2018 amending Directive 2010/31/EU on the energy performance of buildings and Directive 2012/27/EU on energy efficiency. Official Journal of the European Union, 30 May 2018
35. Commission Recommendation (EU) 2019/786, Official Journal of the European Union, 8 May 2019
36. Commission Recommendation (EU) 2019/1019, Official Journal of the European Union, 7 June 2019
37. Stakeholder Consultation on the Renovation Wave Initiative, European Commission, October 2020
38. Delivering the European Green Deal, European Commission, July 2021
39. Directive of the European Parliament and of the Council on the energy performance of buildings, European Commission, 2021
40. Communication from the Commission to the European Parliament, the Council, the European Economic and Social Committee and the Committee of the Regions, European Commission, 2022
41. G. S. O. T. Council, Proposal for a Directive of the European Parliament and of the Council on the energy performance of buildings, Council of the European Union 21 October 2022
42. Amendments adopted by the European Parliament on 14 March 2023 on the proposal for a directive of the European Parliament and of the Council on the energy performance of buildings, 14 March 2023
43. Commission welcomes political agreement on new rules to boost energy performance of buildings across the EU, European Commission, 7 December 2023
44. E. Polychroni, M. Bololia, A. Androutsopoulos, Smart energy performance assessment platform and next generation energy performance certificate of buildings. IOP Conf. Ser. Earth Environ. Sci. **1196**(1), 012006 (2023)
45. Directive 2002/91/EC of the European Parliament and of the Council of 16 December 2002 on the energy performance of buildings," Official Journal of the European Communities, 2002
46. J. Falzon, R.D. Gonzi, S. Grima, E. Vella, Energy performance of buildings for incentivisation in energy-efficient structures: An analysis of secondary data in malta. Processes **12**(5), 874 (2024)
47. Compendium of testing and demonstration board (TDB) meetings, December 2021. https://epanacea.eu/?smd_process_download=1&download_id=2763
48. Report on the selection of the open-source simulation tool and its compliance with ISO 52000 series, May 2021. https://epanacea.eu/?smd_process_download=1&download_id=2668
49. ISO 52016 Energy performance of buildings, 2017
50. ISO 52017-1, 2017
51. SRI Support Team, Waide Strategic Efficiency Europe (Ireland), Research to Market (R2M) Solution (France) and LIST, the Luxembourg Institute of Science and Technology, Smart Readiness Indicator (SRI) Training slide deck, in Version 2.0, January 2022

52. A. Aminlou, M.M. Hayati, K. Zare, P2P Energy Trading in a Community of Individual Consumers with the Presence of Central Shared Battery Energy Storage, in *Demand-Side Peer-to-Peer Energy Trading*, ed. by V. Vahidinasab, B. Mohammadi-Ivatloo, (Springer International Publishing, Cham, 2023), pp. 143–159
53. M.M. Hayati, A. Safari, M. Nazari-Heris, A. Oshnoei, Hydrogen-Incorporated Sector-Coupled Smart Grids: A Systematic Review and Future Concepts, in *Green Hydrogen in Power Systems*, ed. by V. Vahidinasab, B. Mohammadi-Ivatloo, J. Shiun Lim, (Springer International Publishing, Cham, 2024), pp. 25–58
54. H. Majidi-Gharehnaz, H. Biabani, A. Aminlou, M.M. Hayati, M. Abapour, Integration of Solar PV and GH2 in the Future Power Systems, in *Green Hydrogen in Power Systems*, ed. by V. Vahidinasab, B. Mohammadi-Ivatloo, J. Shiun Lim, (Springer International Publishing, Cham, 2024), pp. 203–224
55. A.S. Anđelković et al., Building energy performance certificate—A relevant indicator of actual energy consumption and savings? Energies **14**(12), 3455 (2021)
56. Energy performance certificate.
57. H.S. Fink, Promoting behavioral change towards lower energy consumption in the building sector. Innov. Eur. J. Soc. Sci. Res. **24**(1-2), 7–26 (2011)
58. epccertificates.uk, vol. 7/21/2024, ed
59. L. Yu, S. Qin, M. Zhang, C. Shen, T. Jiang, X. Guan, A review of deep reinforcement learning for smart building energy management. IEEE Internet Things J. **8**(15), 12046–12063 (2021)
60. S. Athanasaki, K. Tsikaloudaki, Smart buildings for smart cities: Analysis of the smart readiness indicator. Green Energy Sustain. **2**(2), 5 (2022)
61. D.A.V. Stijn Verbeke, G. Reynders, Y. Ma, P. Waide, *Final Report on the Technical Support to the Development of a Smart Readiness Indicator for Buildings* (European Commission, Brussels, 2020)
62. G. Plienaitis, M. Daukšys, E. Demetriou, B. Ioannou, P.A. Fokaides, L. Seduikyte, Evaluation of the smart readiness indicator for educational buildings. Buildings **13**(4), 888 (2023)
63. T. Märzinger, D. Österreicher, Extending the application of the smart readiness indicator–A methodology for the quantitative assessment of the load shifting potential of smart districts. Energies **13**(13), 3507 (2020)
64. Supplementing Directive (EU) 2010/31/EU of the European Parliament and of the Council by establishing an optional common European Union scheme for rating the smart readiness of buildings, Official Journal of the European Union, 14 October 2020
65. Directive 2012/27/EU of the European Parliament and of the Council of 25 October 2012 on energy efficiency, amending Directives 2009/125/EC and 2010/30/EU and repealing Directives 2004/8/EC and 2006/32/EC, Official Journal of the European Union, 25 October 2012
66. G. Li et al., A critical review of cyber-physical security for building automation systems. Annu. Rev. Control. **55**, 237–254 (2023)
67. Smart Readiness Indicator, European Commission, December 2022
68. S.V.V. Paul Waide, S. Dourlens, R. Decorme, S.K. List, *Smart Readiness Indicator (SRI), Provisional Guidance on the implementation of the SRI* (European Commission, Brussels, 2023)
69. V. Varsami and E. Burman, An evaluation of the smart readiness indicator proposed for buildings, in Building Simulation 2021 (BS2021), 2021: Proceedings of Building Simulation 2021 Conference, 1–3 September 2021
70. E. Janhunen, L. Pulkka, A. Säynäjoki, S. Junnila, Applicability of the smart readiness indicator for cold climate countries. Buildings **9**(4), 102 (2019)
71. E.B. Vasiliki Varsami, *An Evaluation of the Smart Readiness Indicator Proposed for Buildings* (University College London, London)
72. B. Ramezani, M.G. da Silva, N. Simoes, Application of smart readiness indicator for Mediterranean buildings in retrofitting actions. Energ. Buildings **249**, 111173 (2021)

73. K. Jung, B. Kulvatunyou, S. Choi, M.P. Brundage, An Overview of a Smart Manufacturing System Readiness Assessment, in *Advances in Production Management Systems*, (Springer, Cham, 2016), pp. 705–712
74. V. Eloranta, Applicability evaluation of smart readiness indicator for buildings, 2020
75. The New EN ISO 52120 is replacing EN 15232,EU.BAC guide, European Building Automation Controls Association
76. V. Eloranta, Applicability evaluation of smart readiness indicator for buildings," Master, Lappeenranta–Lahti University of Technology LUT, LUT School of Energy Systems, 2020
77. L. Canale et al., Estimating the smart readiness indicator in the italian residential building stock in different scenarios. Energies **14**(20), 6442 (2021)

Chapter 4
Optimal Operation of Integrated Electricity and Natural Gas Distribution Networks in Smart Cities

Asma Nasiri, Nima Nasiri, Ebrahim Babaei, and Sajad Najafi Ravadanegh

Abbreviations Sets and Indexes

t, s, e	Indices of time, scenario, SPL
k, c, u	Indices of gas well, interruptible load, gas storage
w, v, g	Indices of WT,PV,DG
$m. n$	Indices of NGDN
l, i, j	Indices of line, EDN
Ω_S	Set of scenario
Ω_{DG}	Set of DGs

Parameters

C_t^{in}	Cost of purchased electricity from upstream network ($/MWh)
C_g^{DG}, C_c^{ll}	Power generated cost by DG, interruptible load cost ($/MWh)
C_k^{Gas}	Gas well production cost ($/MWh)
C_u^{ch}, C_u^{dis}	Charge and discharge cost of gas storage system ($/MWh)
C_t^{EV}	Battery destroying cost of the EVs ($/MWh)
η_c, η_e	Charge and discharge efficiency of EVs in SPL (%)
EB_e	Battery capacity of EVs

A. Nasiri · E. Babaei
Faculty of Electrical and Computer Engineering, Energy System Research Institute (ESRI), University of Tabriz, Tabriz, Iran

N. Nasiri (✉) · S. N. Ravadanegh
Resilient Smart Grids Research Lab, Electrical Engineering Department, Azarbaijan Shahid Madani University, Tabriz, Iran

A. Ahmadian et al. (eds.), *Sustainable Energy Resources in Smart Cities*, Green Energy and Technology, https://doi.org/10.1007/978-3-031-91516-1_4

$SOC_{e,s}^{in}$	Initial value of SOC for EV (%)
$SOC_{e}^{\min}$	Minimum limit of SOC (%)
$DT_{e,s}$, $ECPM_{e}$	Daily covered distance of EV, Energy consumption amount of EVs (mile)
$\overline{SOC_{e}}, \underline{SOC_{e}}$	Maximum and minimum amount of SOC in EVs (%)
$SOC_{e,s}^{end}$	Highest amount of SOC in EVs (%)
$SOC_{e,s}^{des}$	Desired amount of SOC in EVs (%)
Cr_{e}	Batteries nominal charging rate of EVs (KW)
BC_{e}	EV battery capacity (KW)
a_{0}	Cycle depth degradation function coefficient in EVs (%)
$MD_{e,s,t}$	Cost of marginal degradation in EVs ($/MWh)
β^{dg}	Calendar degradation coefficient
$T_{e,s,t}^{amb}$	EV battery temperature
$\overline{P_{g}^{DG}}, \underline{P_{g}^{DG}}$	Maximum and minimum active power generation amount of DGs (MW)
$\overline{Q_{g}^{DG}}, \underline{Q_{g}^{DG}}$	Maximum and minimum reactive power generation amount of DGs (MVar)
$\overline{S_{g}^{DG}}$	Maximum limit of DG complex power
R_{g}^{UP}, R_{g}^{DN}	Ramp-up and ramp-down amounts of DGs (MW)
$\overline{P_{t}^{in}}, \underline{P_{t}^{in}}$	Maximum and minimum active power trade-off between END and upstream network (MW)
$\overline{Q_{t}^{in}}, \underline{Q_{t}^{in}}$	Maximum and minimum reactive power trading between END and upstream network (MVar)
$\overline{S_{in}}$	Maximum limit of upstream network complex power
$\overline{P_{w,s}^{WT}}, \overline{P_{v,s}^{PV}}$	Maximum power generation amounts of WT and PV(MW)
$r_{i,j}^{l}, x_{i,j}^{l}$	Magnitude of resistance and inductance of EDN feeders (P.U)
$z_{i,j}^{l}$	Magnitude of impedance of EDN (P.U)
$\overline{G_{k}}, \underline{G_{k}}$	Maximum and minimum amounts of gas wells (KCF)
$\overline{\Pr_{n}}, \underline{\Pr_{n}}$	Maximum and minimum amounts of gas pressure in NGDN (Bar)
$K_{n,m}^{f}$	NGN pipeline fixed coefficient (%)
$\eta_{u}^{ch}, \eta_{u}^{dis}$	Charge and discharge efficacy of gas storage system (%)
$\overline{G_{u}^{GSS}}, \underline{G_{u}^{GSS}}$	Maximum and minimum capacity bounds of gas storage (Kcf)
G_{ut}^{ch}, G_{u}^{dis}	Maximum charge and discharge amounts of gas storage (Kcf)
ϕ	Conversion factor of gas to power in NGDGs (Kcf /MWh)
η_{g}^{DG}	Efficiency of NGDGs (%)

Variables

$P_{s,t}^{in}, Q_{s,t}^{in}$	Active and reactive power trade-off between EDN and upstream network (MW-Mvar)
$P_{g,s,t}^{DG}, Q_{g,s,t}^{DG}$	Active and reactive output power generated by DGs (MW)
$P_{c,t}^{IL}, Q_{c,t}^{IL}$	Active and reactive magnitudes of interruptible load (MW)
$G_{k,t}$	Magnitude of gas wells production (v)
$G_{u,s,t}^{ch}, G_{u,s,t}^{dis}$	Charge and discharge amount of gas storage system (Kcf)
$dg_{e,t}^{NL}$	Cycle degradation of the EVs located in SPLs
$SOC_{e,s,t}$	State of charge of SPLs (%)
$PC_{e,s,t}$	Charge power of EV batteries (KW)
$PD_{e,s,t}$	Discharge power of EV batteries (KW)
$\sigma_{e,s,y}$	Cycle depth related to EVs (%)
$\psi_{e,s,t}$	Battery cycle depth degradation function related to the EVs
$P_{e,s,t}^{dis}, P_{e,s,t}^{ch}$	Discharge and charge amounts of EVs (MW)
$dg_{e,t}^{con}$	Battery calendar degradation of the EVs ($)
$dg_{e,s,t}^{NL}$	Battery cycle degradation of the EVs (%)
$P_{w,s,t}^{WT}, P_{v,s,t}^{PV}$	Output power of WT and PV (MW)
$V_{i,s,t}$	Buses voltage of EDN
$I_{i,j,s,t}^{l}$	Electricity current flow between two buses i,j in EDN (A)
$P_{i,j,s,t}^{l}, Q_{i,j,s,t}^{l}$	Active and reactive power that flows between two buses i,j in EDN (MW and Mvar)
$P_{e,s,t}^{EV}$	Amount of active power trade-off between SPLs and EDN (MW)
$P_{d,s,t}^{load}$	Electrical active power demand (MW)
$G_{w,s,t}^{well}$	Amount of gas production from gas wells (Kcf)
$\mathrm{Pr}_{n,s,t}$	Magnitude of pressure in gas network nodes (Bar)
$G_{u,s,t}^{GSS}$	Level of gas in gas storage system (Kcf)
G_{gl}^{load}	Gas demand (Kcf)
$G_{g,s,t}^{DG}$	Amount of gas that is produced by gas-fired DGs
C_{in}^{t}	Wholesale electricity market price ($/MWh)
$\overline{C_{in}^{t}}$	Maximum price deviation of wholesale electricity market ($/MWh)
$\underline{C_{in}^{t}}$	Minimum price deviation of wholesale electricity market ($/MWh)

4.1 Introduction

Today, even though most countries have a plan to produce electricity from renewable energies, still nearly a quarter of the electricity in the world is generated by gas [1]. In 2023, the share of natural gas in electricity generation of the USA was 43.1% of total power generation. Figure 4.1, shows that the share of gas power generation of the USA increased in 2022. In the USA, over time, as the share of coal in electricity generation has decreased, the share of gas has increased and somehow gas has replaced coal. Figure 4.2 shows the increasing share of natural gas in generating electricity of the USA from the year 2000 to 2023 [2]. From this data, we can understand that the share of gas in 2023 has an increase of 27.1% compared to the year 2000. As we can see, natural gas has a crucial role in energy supply and generation [3]. Thus, it's an obligation to investigate the integration of gas with other energy sources such as electricity.

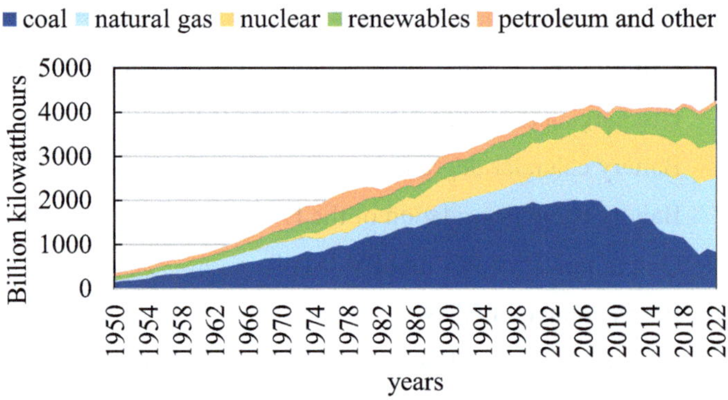

Fig. 4.1 U.S. electricity generation by major energy sources, 1950–2022 [2]

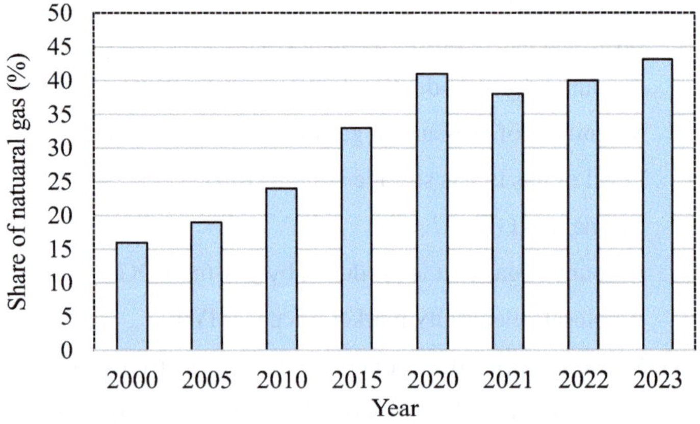

Fig. 4.2 Share of natural gas in electricity generation of the USA [2]

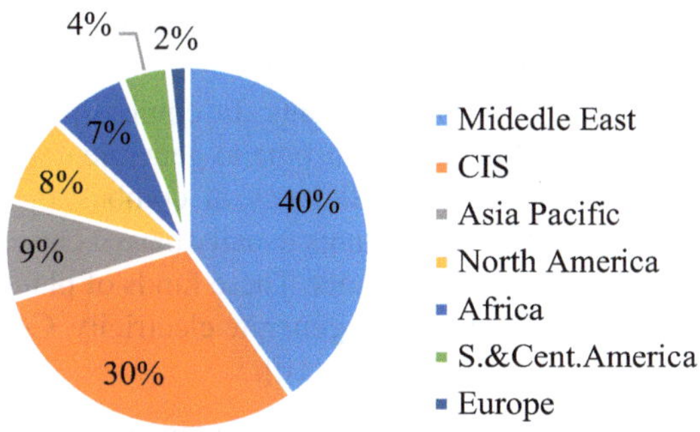

Fig. 4.3 Distribution of proved reserves in 2020 [4]

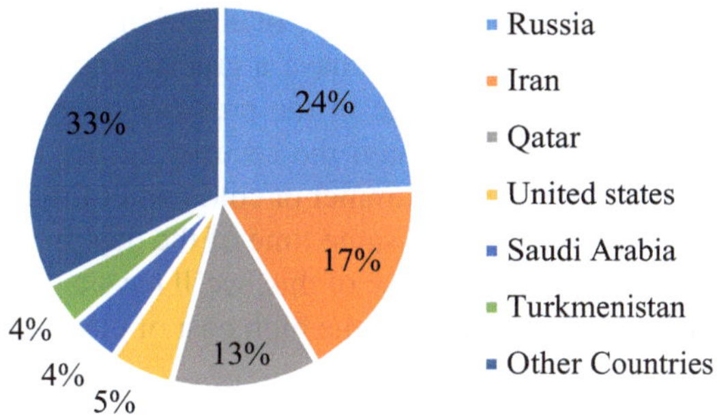

Fig. 4.4 Distribution of proved reserves by countries [5]

There are 188.1 trillion cubic meters of proven gas reserves in the world due to the research in 2020 [4]. Figure 4.3 indicates how natural gas is distributed between different countries of the world. Based on this data, it can be seen that a majority of natural gas is located in the Middle East. Also in this ranking, CIS[1] has second place. As stated in Fig. 4.4, proven gas reserves are dominated by countries of Russia, Iran, and Qatar [5].

Natural gas has many advantages that make it a good choice to be used as fuel to generate electricity. Natural gas has low cost and high efficiency and can be easily stored and delivered through pipelines, and because of these aspects, it had rapid growth in the world [6]. Since natural gas has the least amount of carbon emission among fossil fuels, we can use gas electricity generation as a backup for variable renewable energy in the future. Also, gas-fired units can be controlled easily, and units can be quickly turned off or on (i.e. these plants have low start-up and shut-down times; also they have a fast ramp-down and ramp-up) [7]. For example, 2–10 h are needed for the full operation of coal-fired units, while gas-fired units have a startup time of 10 min to operate [8, 9].

We can sort the gas-fired power plants into three categories [10]:

1. Steam power plants: In a steam power plant, the gas is fired to boil the water and then turn it into steam. The created steam turns the turbine which is coupled with

[1] The commonwealth of independent states.

the rotor of the generator and generates electricity. These kinds of plants usually have efficiencies of 30–35%.

2. Combustion power plants: In combustion power plants natural gas is burnt to turn the blades of a turbine to generate electricity with the generator coupled to the turbine. The efficiency of combustion power plants is between 35% and 40%. Also, these plants usually have a capacity of 100–400 MW.
3. Combined cycle plants: These kinds of plants use both gas and steam to turn the blade of turbine and generate electricity. Combined cycle plants usually have an efficiency of 50–60% [10].

In integrated gas and electric networks, the economics and security of the power grid can be substantially impacted due to the interrelationship of gas and electricity [11]. Due to the dependence of electricity generation on gas-fired units, any change in the price of gas will have a crucial impact on the cost of electricity. Also, any interruption or change in the pressure of gas in pipelines would drop the power generation and threaten the security of the power grid. In some special weather conditions like hot summer or cold winter times, we confront a high demand for gas and electricity at the same time of the day. In these conditions, the energy cost will get higher, and because of this, we'll have a growth in the cost of gas-fired units. As a result, we'll confront the high cost of electricity in the market [12].

4.2 Literature Review

A scheduling of distribution feeder reconfiguration (DFR) and microgrids is stated in [13] with taking into account the uncertainties associated with renewable energies, electricity demand, and upstream market prices. A comprehensive study of distributed energy resources (DERs) on distribution networks is presented in [14]. This paper studies the impact of DERs on distribution networks and shows how to model the DERs by deterministic and probabilistic approaches in power systems. Paper [15] presented an approach for the integration of an incentive-based demand response (DR) program and energy management of microgrids on a day-ahead time framework by considering the uncertainties in renewable energy resources. A two-stage frame of decentralized scheduling of microgrids is presented in [16]. These stages include solving the DFR and day-ahead scheduling problems. A two-stage stochastic approach is proposed in [17] to investigate the impact of integrating renewable energy resources with energy storage systems in the presence of extensive load aggregators by considering uncertainty and using DR programs. Review paper [18] studies the classification of distributed generation (DG) and the challenges related to integration of DGs into distribution networks. Additionally, it investigates the experiences of countries that have coordination of DG and grid in their power system. An investigation of decentralized energy management for an active distribution system (ADN) is stated in [19]. This paper considers the participation role of aggregated office buildings in the optimization of an ADN.

A comprehensive review of distributed optimization for electric distribution networks is presented in [20]. This review paper proposes the different types of problems and introduces an algorithm to solve all of them separately. Paper [21] provided a study of energy management in electric vehicles (EVs) and showed how to solve these problems with convex optimization methods. Article [22] investigates different optimization strategies for optimization of distribution networks with distributed generation and EVs. An energy management strategy of grid-connected distribution systems is presented in [23] considering the high penetration of EVs. An exhaustive investigation of how to model the distribution system and all related components (i.e. EVs, renewable energies, etc.) is stated in [24]. This article also shows how to use different types of optimization methods. Article [25] proposes a management method for decreasing the impact of the high penetration of EVs into the distribution network.

A coordinated energy system optimal dispatching technique with taking into consideration the uncertainty and power to gas (P2G) units is presented in [26]. Paper [27] provides an optimal scheduling of coordinated electricity and gas distribution networks under the uncertainties by considering the power to gas (P2G) units. A two-stage optimal model of an integrated natural electricity and gas network is provided in [28]. This article uses a stochastic approach to model the uncertainty of renewable energy generation and gas and electricity demands. Article [29] presented an optimization model of coordinated gas and electric distribution networks in the existence of P2G units, DGs, demand side management (DSM), and energy storages.

Some studies, such as [13–25], have focused on the optimization of electrical distribution networks (EDNs) in the company of gas-fired distributed generations (GFDG) and non-gas-fired distributed generations (NGDG). However, they have ignored the impact of natural gas distribution network (NGDN) constraints on the operation of the EDN. The high penetration of GFDG in EDNs can significantly affect the NGDN operation. Therefore, ignoring the constraints of the NGDN leads to unrealistic results.

Some researches such as [13–20, 26–28] have ignored the impact of electric parking lots in coordinated NGDNs and EDNs. The high penetration of EVs in EDNs leads to the creation of new peak hours, which must be effectively addressed. Ignoring the high penetration of EVs in future distribution networks will lead to significant challenges such as lack of production, high losses, and instability in EDNs.

Mathematical modeling presented in studies [13–19, 26–29] are complex and non-convex. Modeling the mathematical programming problem within the framework of convex optimization can significantly enhance scalability and reduce problem-solving time. Hence, the use of convex optimization makes it possible to quickly solve large-scale power system models in very short time frames. The solution obtained from the structure of the convex optimization model represents a global optimal solution, thus serving as a strong confirmation of the validity of the results obtained from the simulations.

Table 4.1 Comparison of this research with prior studies

References	Electricity network	Renewable Energies	Uncertainty	Convex pptimization	Smart parking lot	Gas network	Gas storage
[13]	✓	✓	✓	×	×	×	×
[14]	✓	✓	✓	×	×	×	×
[15]	✓	✓	✓	×	×	×	×
[16]	✓	✓	✓	×	×	×	×
[17]	✓	✓	✓	×	×	×	×
[18]	✓	✓	×	×	×	×	×
[19]	✓	✓	×	×	×	×	×
[20]	✓	×	×	✓	×	×	×
[21]	✓	×	×	✓	✓	×	×
[22]	✓	✓	✓	✓	✓	×	×
[23]	✓	✓	✓	✓	✓	×	×
[24]	✓	✓	✓	✓	✓	×	×
[25]	✓	✓	✓	✓	✓	×	×
[26]	✓	✓	✓	×	×	✓	×
[27]	✓	✓	✓	×	×	✓	✓
[28]	✓	✓	✓	×	×	✓	✓
[29]	✓	✓	✓	×	✓	✓	✓

Based on the gaps identified in the above researches and Table 4.1, this research makes the following contributions:

1. A hybrid stochastic-robust programming based on convex optimization is proposed for integrated natural gas and electricity systems in the company of smart parking lots and gas storage systems.
2. The impact of the smart charging strategy of EV implementation on the technical and economic performance of coordinated electricity and natural gas distribution systems has been effectively examined.
3. The advantage of integrating gas storage systems in the natural gas distribution system has been studied on increasing fuel supply to industrial gas loads such as GFDG and improving the operational and economical efficiency of the electricity distribution system.

4.3 Description Problem

As previously mentioned, studying the interdependency of gas and electric networks is crucial for operators, system designers, and stakeholders who are responsible for ensuring the energy security of networks [10]. Figure 4.5 shows the illustration of an integrated gas and electrical system coupled by GFDG. We presume that the

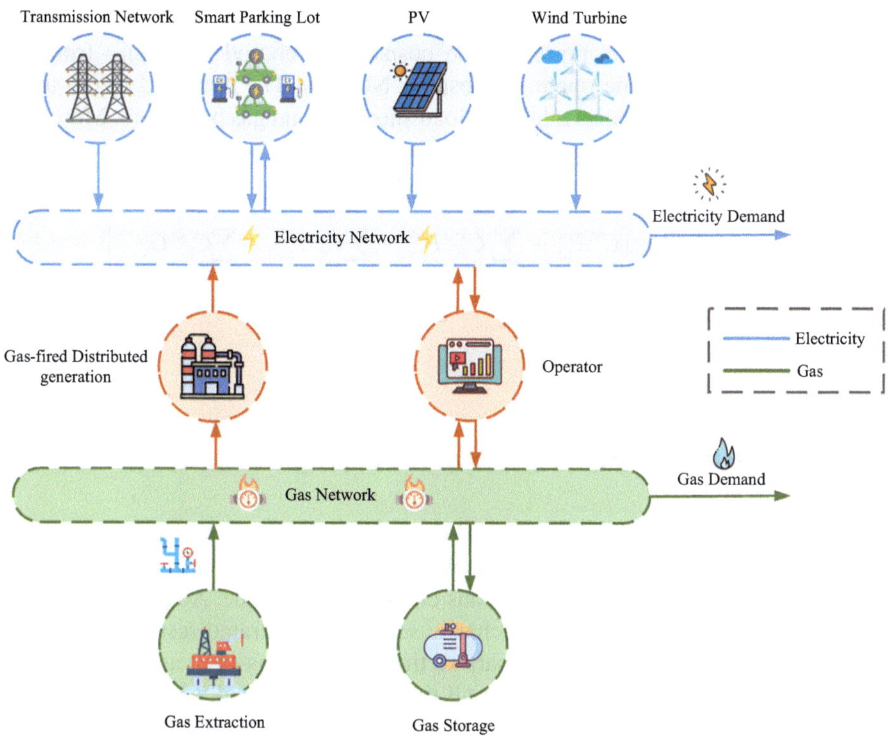

Fig. 4.5 Diagram of integrated gas and electricity networks

distribution network is the owner of the smart parking lot, photovoltaic (PV), and wind turbine. In the electrical network, we have upstream network (transmission network), PV, and wind turbines as suppliers of electricity while electricity demand represents the consumer of the distribution system. Additionally, the smart parking lot has the role of both consumer and supplier of electricity.

In the NGDN, we have gas extraction as a supplier of gas while the gas demand represents the consumption side. In this system, gas storage can play both roles as a consumer and supplier. In gas peak demand, the storage injects gas into the network and stores it at off-peak times (i.e., such as at night).

4.4 Mathematical Formulation

4.4.1 Objective Function

Equation (4.1) displays the objective function of the proposed approach. As is well known, the main purpose of the proposed problem is to minimize costs related to the operation of the integrated urban gas and electricity distribution networks

(IUEAGDNs). The modeled objective function in Eq. (4.1) comprises six terms in which the first two terms represent the costs of purchased electricity from the upstream network and the operating costs of NGDNs. The third and fourth terms include the modeling of costs related to load shedding and gas producers. Moreover, the operating costs of the gas storage system (GSS) and the cost of destroying the EV batteries are stated in the fifth and sixth terms.

$$OF_1 = \min \sum_{s\in\Omega_s} \pi_s \sum_t \left\{ \begin{array}{l} C_t^{in} P_{t,s}^{in} + \sum_{g\in\Omega_{cu}} C_g^{DG} P_{g,s,t}^{DG} + \sum_c C_c^{IL} P_{c,s,t}^{IL} + \sum_k C_k^{Gas} G_{k,s,t} \\ + \sum_u \left(C_u^{ch} G_{u,s,t}^{ch} + C_u^{dis} G_{u,s,t}^{dis} \right) + \sum_e dg_{e,t}^{con} \end{array} \right\} \tag{4.1}$$

4.4.2 Constraints Related to Smart Parking Lot

In the presented approach, all smart parking lots (SPLs) are modeled in the form of a convex space by considering the probability distribution function of vehicle owners' behavior, with applying the k-means clustering technique. Additionally, all the EVs are modeled by a series of stochastic scenarios generated rest on daily miles driven as well as vehicle arrival/departure times. Equations (4.2) and (4.3) express the state of charge (SOC) of SPLs for all time horizons as well as arrival/departure times. The initial SOC of SPLs is represented by Eq. (4.4), and the final SOC of SPLs at the time of vehicle departure is demonstrated between the lowest desired level of SOC and the highest possible level of SOC. The maximum/minimum limitation of EV batteries is also modeled through Eq. (4.6). Equations (4.7)–(4.9) utilize two variables, namely, $P_{e,s,t}^{ch}$ (free variable in sign) and $P_{e,s,t}^{dis}$ (negative variable) to separate discharge and charge power. The depth of the battery cycle is expressed by Eq. (4.10). Limitation (4.11) expresses the final destruction of EV batteries based on KW. Finally, the overall battery wear of SPLs is modeled through Eq. (4.12)[30].

$$SOC_{e,s,t} = SOC_{e,s,t\ 1} + \frac{\eta_c . P_{e,s,t}^{ch}}{EB_e} - \frac{P_{e,s,t}^{dis}}{EB_e . \eta_e} \quad \forall e,s,t : \{ t | t \in t \neq TA_e \} \tag{4.2}$$

$$SOC_{e,s,t} = SOC_{e,s}^{in} + \frac{\eta_c . P_{e,s,t}^{ch}}{EB_e} - \frac{P_{e,s,t}^{dis}}{EB_e . \eta_e} \quad \forall e,s,t : \{ t | t \in t = TA_e \} \tag{4.3}$$

$$SOC_{e,s}^{In} = \max\left(SOC_e^{min}, 1 - \frac{DT_{e,s} \times ECPM_e}{EB_e} \right) \quad \forall e,s \tag{4.4}$$

$$SOC_{e,s,t} = \min\left(SOC_{e,s}^{end}, SOC_e^{des} \right) \quad \forall e,s,t : \{ t | t \in t = TD_e \} \tag{4.5}$$

$$\underline{SOC_e} \le SOC_{e,s,t} \le \overline{SOC_e} \quad \forall e,s,t \tag{4.6}$$

$$-Cr_e \leq P^{ch}_{e,s,t} \leq Cr_e \quad \forall e,s,t \tag{4.7}$$

$$-Cr_e \leq P^{dis}_{e,s,t} \leq P^{ch}_{e,s,t} \quad \forall e,s,t \tag{4.8}$$

$$P^{dis}_{e,s,t} \leq 0 \quad \forall e,s,t \tag{4.9}$$

$$\sigma_{e,s,t} = \frac{P^{dis}_{e,s,t}}{EB_e.\eta_e} - \sigma_{e,s,t\ 1} \quad \forall e,s,t \tag{4.10}$$

$$\psi_{e,s,t}\left(\sigma_{e,s,t}\right) = a_0 \cdot \left(\sigma_{e,s,t}\right)^{2.03} \quad \forall e,s,t \tag{4.11}$$

$$\begin{aligned} MD_{e,s,t} &= BC_e \cdot \frac{\partial \psi_{e,s,t}}{\partial P^{dis}_{e,s,t}} = BC_e \cdot \frac{\partial \psi_{e,s,t}}{\sigma_{e,s,t}} \cdot \frac{\sigma_{e,s,t}}{\partial P^{dis}_{e,s,t}} \\ &= 2.03 a_0 \cdot \frac{BC_e}{EB_e.\eta_e} . \sigma_{e,s,t}{}^{1.03} \quad \forall e,s,t \end{aligned} \tag{4.12}$$

$$dg^{NL}_{e,s,t} = P^{dis}_{e,s,t} . MD_{e,s,t} \quad \forall e,s,t \tag{4.13}$$

According to Eq. (4.14), the annual battery degradation model has a direct relationship with the temperature and battery charge state. As you can see, the annual battery degradation of SPLs is supposed as convex quadratic equation, that can be converted into a SOCP[2] model as Eq. (4.15) by presuming positive matrix-specific coefficients. The optimal values for Eqs. (4.14) and (4.15) are identical, with the scalar value of $\left(\beta^{dg}_0 + \beta^{dg}_1 T^{amb}_{e,t} + \beta^{dg}_4 \left(T^{amb}_{e,t}\right)^2 - \frac{1}{4}\left(\beta^{dg}_2 + \beta^{dg}_3 T^{amb}_{e,t}\right)\left(\beta^{dg}_5\right)^{-1} \right)$ should be added to the objective function value [30].

$$\begin{aligned} dg^{cal}_{e,s,t} &= \beta^{dg}_0 + \beta^{dg}_1 T^{amb}_{e,s,t} + \beta^{dg}_2 SOC_{e,s,t} + \beta^{dg}_3 T^{amb}_{e,s,t} SOC_{e,s,t} \\ &+ \beta^{dg}_4 \left(T^{amb}_{e,s,t}\right)^2 + \beta^{dg}_5 \left(SOC_{e,s,t}\right)^T \left(SOC_{e,s,t}\right) \forall e,t \end{aligned} \tag{4.14}$$

$$\left\| \begin{matrix} \left(\beta^{dg}_5\right)^{0.5} SOC_{e,s,t} \\ \frac{1}{2}\left(\beta^{dg}_5\right)^{-0.5} \left(\beta^{dg}_2 + \beta^{dg}_3 T^{amb}_{e,s,t}\right)^T \end{matrix} \right\|_2 \leq dg^{con}_{e,t} \tag{4.15}$$

4.4.3 Constraints Related to DGs

Generally, the constraints of DGs are modeled by Eqs. (4.16)–(4.25). Equations (4.16)–(4.18) model active, reactive, and complex generation power constraints, respectively. Ramp-up and ramp-down constraints are also modeled by Eqs. (4.19)

---

[2] Second-order conic programming.

and (4.20), respectively. Equations (4.21)–(4.23) also model the constraints of active, reactive, and complex power purchased from the upstream network, respectively. Finally, the forecasted maximum power constraints of wind turbine and photovoltaic system are modeled by Eqs. (4.24) and (4.25), respectively.

$$\underline{P_{g}^{DG}} \leq P_{g,s,t}^{DG} \leq \overline{P_{g}^{GD}} \quad \forall g \notin \Omega_{DG}, \forall s, \forall t \tag{4.16}$$

$$\underline{Q_{g}^{DG}} \leq Q_{g,s,t}^{DG} \leq \overline{Q_{g}^{GD}} \quad \forall g \notin \Omega_{DG}, \forall s, \forall t \tag{4.17}$$

$$\begin{matrix} P_{g,s,t}^{DG} \\ Q_{g,s,t}^{DG} \end{matrix} \leq \overline{S_{g}^{DG}} \quad \forall g \in \Omega_{DG}, \forall s, \forall t \tag{4.18}$$

$$P_{g,s,t}^{DG} - P_{g,s,t\;1}^{DG} \leq R_{g}^{UP} \quad \forall g, \forall s, \forall t \tag{4.19}$$

$$P_{g,s,t\;1}^{DG} - P_{g,s,t}^{DG} \leq R_{g}^{DN} \quad \forall g, \forall s, \forall t \tag{4.20}$$

$$\underline{P_{t}^{in}} \leq P_{s,t}^{in} \leq \overline{P_{t}^{in}} \quad \forall s, \forall t \tag{4.21}$$

$$\underline{Q_{t}^{in}} \leq Q_{s,t}^{in} \leq \overline{Q_{t}^{in}} \quad \forall s, \forall t \tag{4.22}$$

$$\begin{matrix} P_{s,t}^{in} \\ Q_{s,t}^{in} \end{matrix} \leq \overline{S_{s,t}^{in}} \quad \forall s, \forall t \tag{4.23}$$

$$0 \leq P_{w,s,t}^{WT} \leq \overline{P_{w,s}^{WT}} \quad \forall w, \forall s, \forall t \tag{4.24}$$

$$0 \leq P_{v,s,t}^{PV} \leq \overline{P_{v,s}^{PV}} \quad \forall v, \forall s, \forall t \tag{4.25}$$

4.4.4 Electric Distribution Network (EDN) Constraints

In this study, the EDN is modeled using Eqs. (4.26)–(4.33). Constraint (4.29) is the squared voltage modeler for the convex modeling of the EDN. The current flow through the EDN feeders is modeled by (4.27). The constraints (4.28) and (4.29) represent the modeling of the active and reactive power balance equations within the EDN. The constraint referred to the voltage of EDN nodes is also expressed by (4.30). In order to take advantage of convex optimization, the nonlinear equation (4.27) is relaxed in the form of constraint (4.32). Equation (4.33) represents the constraint on the power flow through the feeders of the EDN [31].

$$U_{i,s,t} = V_{i,s,t}^{2} \quad \forall i, \forall s, \forall t \tag{4.26}$$

$$I_{i,j,s,t}^{l} = \left(P_{i,j,s,t}^{l}\right)^{2} + \left(Q_{i,j,s,t}^{l}\right)^{2} / V_{i,j,s,t}^{2} \quad \forall l, \forall s, \forall t \tag{4.27}$$

$$
\begin{aligned}
&P^{l}_{i,j,s,t} - r^{l}_{i,j} I^{l}_{i,j,s,t} + \sum_{w \in A^{w}_{j}} P^{\mathrm{WT}}_{v,s,t} + \sum_{v \in A^{v}_{j}} P^{\mathrm{PV}}_{v,s,t} + \sum_{g \in A^{g}_{j}} P^{\mathrm{DG}}_{g,s,t} = \\
&\sum_{d \in A^{d}_{j}} P^{\mathrm{Load}}_{d,s,t} + \sum_{e \in A^{e}_{j}} P^{\mathrm{EV}}_{e,s,t} + \sum_{k \in A^{k}_{j}} P^{l}_{j,k,s,t} : \lambda^{\mathrm{DLMP}}_{i,s,t} \quad \forall i, \forall s, \forall t
\end{aligned} \tag{4.28}
$$

$$
Q^{l}_{i,j,s,t} - x^{l}_{i,j} I^{l}_{i,j,s,t} + \sum_{g \in A^{g}_{j}} Q^{\mathrm{DG}}_{g,s,t} = \sum_{d \in A^{d}_{j}} Q^{\mathrm{Load}}_{d,s,t} + \sum_{k \in A^{k}_{j}} Q^{l}_{j,k,s,t} \quad \forall i, \forall s, \forall t \tag{4.29}
$$

$$
U_{j,s,t} = U_{i,s,t} - 2\left(P^{l}_{i,j,s,t} r^{l}_{i,j} + Q^{l}_{i,j,s,t} x^{l}_{i,j}\right) + \left(z^{l}_{i,j}\right)^{2} I^{l}_{i,j,s,t} \quad \forall l, \forall s, \forall t \tag{4.30}
$$

$$
\underline{U_{j}} \le U_{j,s,t} \le \overline{U_{j}} \quad \forall j, \forall s, \forall t \tag{4.31}
$$

$$
\left\| \begin{matrix} 2P^{\mathrm{l}}_{\mathrm{i,j,s,t}} \\ 2Q^{\mathrm{l}}_{\mathrm{i,j,s,t}} \\ I^{\mathrm{I}}_{\mathrm{i,j,s,t}} - U_{\mathrm{i,s,t}} \end{matrix} \right\|_{2} \le I^{\mathrm{I}}_{\mathrm{i,j,s,t}} - U_{\mathrm{i,s,t}} \quad \forall \mathrm{l}, \forall \mathrm{s}, \forall \mathrm{t} \tag{4.32}
$$

$$
\left\| \begin{matrix} P^{l}_{i,j,s,t} \\ Q^{l}_{i,j,s,t} \end{matrix} \right\|_{2} \le S_{l} \quad \forall j, \forall s, \forall t \tag{4.33}
$$

4.4.5 *Constraints of Natural Gas Distribution Network (NGDN)*

Generally, the constraints of NGDN are presented by Eqs. (4.34)–(4.54). Equations (4.34) and (4.35) are the constraints of gas wells and the allowable pressure value in gas network nodes, respectively. The natural gas flow in the pipeline of the gas network is presented as a nonlinear equation (4.36), which is relaxed by Eq. (4.37). Equations (4.38) and (4.39) express the GSS at time $t > 1$ and $t = 1$, respectively. The maximum/minimum storage limit of GSS is also modeled by Eq. (4.40). Equations (4.41) and (4.42) model the charge and discharge limitations of GSS, respectively. The Eq. (4.43) shows gas consumption of GFDGs. The gas fuel balance equation is also modeled by constraint (4.44) [32].

$$
\underline{G}_{k} \le G^{\mathrm{well}}_{k,s,t} \le \overline{G}_{k} \quad \forall k, \forall s, \forall t \tag{4.34}
$$

$$
\underline{\mathrm{Pr}}_{n} \le \mathrm{Pr}_{n,s,t} \le \overline{\mathrm{Pr}}_{n} \quad \forall n, \forall s, \forall t \tag{4.35}
$$

$$
\left(G^{\mathrm{flow}}_{n,m,s,t}\right)^{2} = K^{f}_{n,m}\left(\mathrm{Pr}^{2}_{n,s,t} - \mathrm{Pr}^{2}_{m,s,t}\right) \quad \forall n, \forall m, \forall s, \forall t \tag{4.36}
$$

$$
\left\| \begin{matrix} G^{\mathrm{flow}}_{n,m,s,t} \\ K^{f}_{n,m} \mathrm{Pr}_{m,s,t} \end{matrix} \right\|_{2} \le K^{f}_{n,m} \mathrm{Pr}_{n,s,t} \quad \forall n, \forall m, \forall s, \forall t \tag{4.37}
$$

$$G_{u,s,t}^{\text{GSS}} = G_{u,s,t-1}^{\text{GSS}} + G_{u,s,t}^{\text{ch}} \eta_u^{\text{ch}} - \frac{G_{u,s,t}^{\text{dis}}}{\eta_u^{\text{dis}}} \quad \forall u, \forall s, \forall t \geq 1 \tag{4.38}$$

$$G_{u,s,t}^{\text{GSS}} = G_{u,s,ini}^{\text{GSS}} + G_{u,s,t}^{\text{ch}} \eta_u^{\text{ch}} - \frac{G_{u,s,t}^{\text{dis}}}{\eta_u^{\text{dis}}} \quad \forall u, \forall s, \forall t = 1 \tag{4.39}$$

$$\underline{G_u^{\text{GSS}}} \leq G_{u,s,t}^{\text{GSS}} \leq \overline{G_u^{\text{GSS}}} \quad \forall u, \forall s, \forall t \tag{4.40}$$

$$0 \leq G_{u,s,t}^{\text{ch}} \leq \overline{G_u^{\text{ch}}} \quad \forall u, \forall s, \forall t \tag{4.41}$$

$$0 \leq G_{u,s,t}^{\text{dis}} \leq \overline{G_u^{\text{dis}}} \quad \forall u, \forall s, \forall t \tag{4.42}$$

$$G_{g,s,t}^{\text{DG}} = \frac{\phi P_{g,s,t}^{\text{DG}}}{\eta_g^{\text{DG}}} \quad \forall g, \forall s, \forall t \tag{4.43}$$

$$\sum_{k \in A_n^k} G_{k,s,t}^{\text{well}} - \sum_{gl \in A_n^{gl}} G_{gl,s,t}^{\text{Load}} - \sum_{g \in A_n^g} G_{g,s,t}^{\text{DG}} = \sum_{u \in A_n^u} \left(G_{u,s,t}^{\text{ch}} - G_{u,s,t}^{\text{dis}} \right) \quad \forall n, \forall s, \forall t \tag{4.44}$$

4.5 Robust Optimization

In this study, the fluctuating price behavior of the wholesale electricity market has been modeled using the robust optimization method. In the robust modeling, unlike stochastic optimization methods, there is no need for different probability distributions. For this purpose, the robust optimization method employs the technique of interval changes $\underline{C}_{\text{t}}^{\text{DS}}$ " C_{t}^{DS} " $\overline{C}_{\text{t}}^{\text{DS}}$ in uncertainty parameters. When predictive models are not applicable, these intervals can be specified as a percentage around the predicted value, encompassing a range of uncertainty. In this case we will have [33]:

$$\overline{C}_t^{\text{in}} = (1 + \alpha) C_t^{\text{in}} \quad \forall t \tag{4.45}$$

$$\underline{C}_t^{\text{in}} = (1 - \alpha) C_t^{\text{in}} \quad \forall t \tag{4.46}$$

The value of α can be adjusted within the range of zero to one to set the desired level of uncertainty value. To model the robust optimization, despite the objective function (4.1), we replace the min and max function (4.47).

$$\min \sum_{s \in \Omega_s} \pi_s \sum_t \left\{ \begin{array}{l} C_t^{\text{in}} P_{t,s}^{\text{in}} + \max\limits_{\{S_0 \| S_0 | \leq \Gamma\}} \left[\sum\limits_{t \in S_0} \overline{C}_t^{\text{in}} - \underline{C}_t^{\text{in}} \right] P_{t,s}^{\text{in}} + \sum\limits_{g \in CU} C_g^{\text{DG}} P_{g,s,t}^{\text{DG}} + \sum\limits_c C_c^{\text{IL}} P_{c,t,s}^{\text{IL}} \\ + \sum\limits_u \left(C_u^{\text{ch}} G_{u,s,t}^{\text{ch}} + C_u^{\text{dis}} G_{u,s,t}^{\text{dis}} \right) + \sum\limits_k C_k^{\text{Gas}} G_{k,s,t} + \sum\limits_e C_e^{\text{EV}} dg_{e,s,t}^{\text{NL}} \end{array} \right\} \tag{4.47}$$

In Eq. (4.47), the value of Γ indicates the level of robustness of the presented model to the level of conservatism of the results. In robust optimization, if Γ = 0, the impact of uncertainty on the price deviation of the electricity wholesale market is completely ignored. On the other hand, if Γ = T, the price uncertainty is regarded in the entire time horizon and the most conservative answer is acquired. Incremental change in the amount of Γ leads to an increase in the level of robustness of the results against any increase in the electricity wholesale market price. As the value of Γ increases, the robustness of the results against fluctuations in the electricity wholesale market price also increases. With the introduction of auxiliary variables $y_{t,s}$ and $z_{t,s}$, the objective function (4.47) can be converted into the function (4.48) [33].

$$\min \sum_{s} \pi_s \sum_{t} \left\{ \begin{array}{l} C_t^{\text{in}} P_{t,s}^{\text{in}} + \max\limits_{\left\{ \sum_{t,s} Z_{t,s} \le \Gamma, 0 \le Z_{t,s} \le 1, P_{t,s}^{in} \le y_{t,s} \right\}} \left[\sum_{t \in S_0} \left(\bar{C}_t^{\text{in}} - \underline{C}_t^{\text{in}} \right) \right] y_{t,s} z_{t,s} \\ + \sum_{g \in CU} C_g^{\text{DG}} P_{g,s,t}^{\text{DG}} \\ + \sum_{c} C_c^{\text{IL}} P_{c,t,s}^{\text{IL}} + \sum_{u} \left(C_u^{\text{ch}} G_{u,s,t}^{\text{ch}} + C_u^{\text{dis}} G_{u,s,t}^{\text{dis}} \right) + \sum_{k} C_k^{\text{Gas}} G_{k,s,t} \\ + \sum_{e} C_e^{\text{EV}} dg_{e,s,t}^{\text{NL}} \end{array} \right\} \tag{4.48}$$

As it is known, the Eq. (4.48) is a bi-level min–max approach and cannot be solved using commercial solvers under normal conditions. Therefore, using the dual theory, Eq. (4.48) is changed into a single-level min problem and replaced by Eqs. (4.49)–(4.54). In these equations, variables $\zeta_{t,s}$ and ψ are used as dual variables to define uncertainty intervals [33].

$$\min \sum_{s \in \Omega} \pi_s \sum_{t} \left\{ \begin{array}{l} C_t^{\text{in}} P_{t,s}^{\text{in}} + \psi.\Gamma + \sum_{t} \zeta_t + \sum_{g \in CU} C_g^{\text{DG}} P_{g,s,t}^{\text{DG}} + \sum_{c} C_c^{\text{IL}} P_{c,t,s}^{\text{IL}} + \\ \sum_{u} \left(C_u^{\text{ch}} G_{u,s,t}^{\text{ch}} + C_u^{\text{dis}} G_{u,s,t}^{\text{dis}} \right) + \sum_{k} C_k^{\text{Gas}} G_{k,s,t} + \sum_{e} C_e^{\text{EV}} dg_{e,s,t}^{\text{NL}} \end{array} \right\} \tag{4.49}$$

$$\psi + \zeta_{t,s} \le \left(\bar{C}_t^{\text{in}} - \underline{C}_t^{\text{in}} \right) y_{t,s} \quad \forall s, \forall t \tag{4.50}$$

$$\zeta_{t,s} \ge 0 \quad \forall s, \forall t \tag{4.51}$$

$$y_{t,s} \ge 0 \quad \forall s, \forall t \tag{4.52}$$

$$P_{t,s}^{\text{in}} \le y_{t,s} \quad \forall s, \forall t \tag{4.53}$$

$$\psi \ge 0 \tag{4.54}$$

4.6 Simulation and Case Studies

The topology of IUEAGDNs is shown in Fig. 4.6. As it is known, an IEEE 33-bus standard test system has been chosen to model the EDN, whose information and data can be extracted by [31]. Moreover, the NGDN is modeled using a standard 20-node system test, whose data is available in [34]. Overall, the EDN has three GFDGs and a NGDG. The location of DGs is clearly defined in Fig. 4.6. Additionally, two wind turbines and two photovoltaic systems are situated at buses 9, 30, 17, and 26. The stochastic production power of wind turbines and photovoltaic systems is shown in Figs. 4.7 and 4.8, respectively. A stochastic programming method is used for modeling the unpredictable behavior of renewable energy sources. To investigate the behavior of EVs in the EDN, three SPLs have been incorporated, and their locations

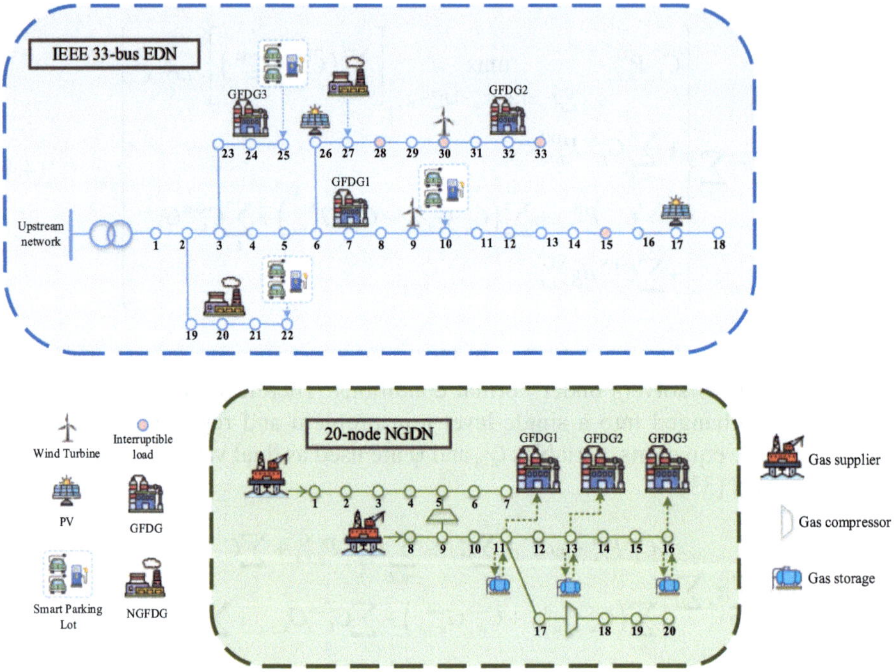

Fig. 4.6 The topology of IUEAGDNs

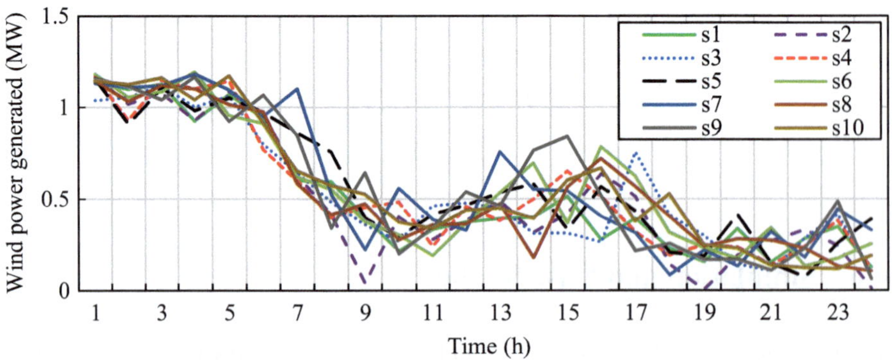

Fig. 4.7 The stochastic production power of wind turbines

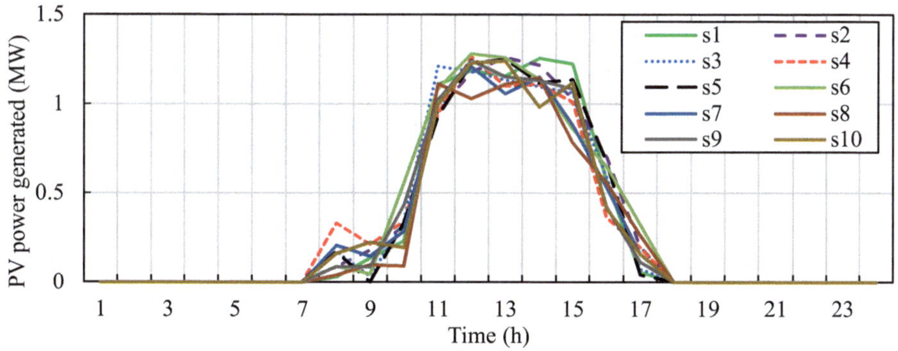

Fig. 4.8 The stochastic production power of PV

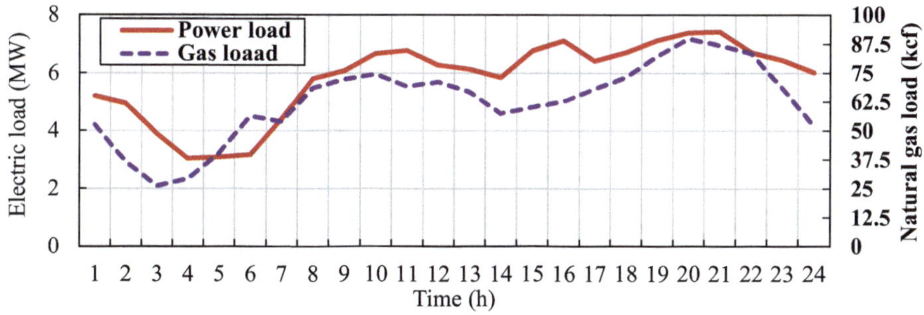

Fig. 4.9 Electricity and natural gas loads

are situated in the 10th, 22th, and 25th buses, respectively. It is assumed that the capacity of each SPL is 1000 EVs. Similarly, the uncertain behavior of EV owners is modeled using stochastic programming methodology. The details of stochastic modeling of EVs using the corresponding probability distributions are explained by [32]. It is worth noting that the data related to EVs, distributed generations (DGs), GSS, and other parameters of the problem is presented in Appendix. The power consumption of NGDN and EDN is illustrated in Fig. 4.9. It should be mentioned that the fluctuating price behavior of the electricity wholesale market is modeled utilizing the robust optimization method. Figure 4.10 displays the price interval within the electricity wholesale market. The effectiveness of the presented approach has been examined in the configuration of three case studies; the details of them are as follows:

1. Optimal stochastic programming of IUEAGDNs in the presence of SPL considering uncontrolled charging strategy
2. Examining the effect of incorporating GSSs on the optimal decision-making of the IUEAGDNs considering the uncontrolled charging strategy
3. Investigating the effect of implementing smart charge strategies on the optimal decision-making of the IUEAGDNs, excluding considerations of GSSs.

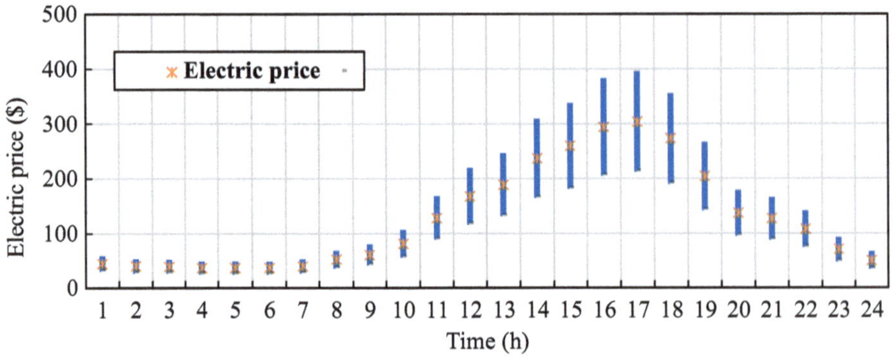

Fig. 4.10 Price interval within the electricity wholesale market

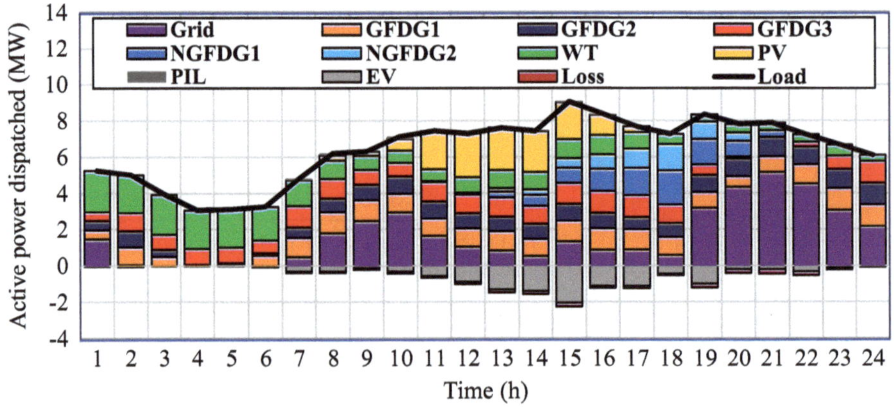

Fig. 4.11 Electrical load supply's optimum scheduling in the EDN

4. Hybrid stochastic-robust scheduling for IUEAGDNs in the company of SPL, considering smart charging strategy and GSS

4.6.1 Case Study 1

In case study 1, a stochastic optimal scheduling is proposed for IUEAGDNs in the presence of SPL, without considering the uncontrolled charging strategy. Figure 4.11 depicts the optimum scheduling of the active consumption power supply of the urban EDN. It is worth noting that the upstream network supplies 25.5% of the electric loads of EDN. Additionally, GFDGs and NGFDGs provide 38.63% and 9.67% of the electric loads of the upstream EDN, respectively. Furthermore, 17.03% and 8.88% of the electric loads of the urban EDN are provided by wind turbine and PV, respectively. Finally, 0.27% of electric loads remain unserviced.

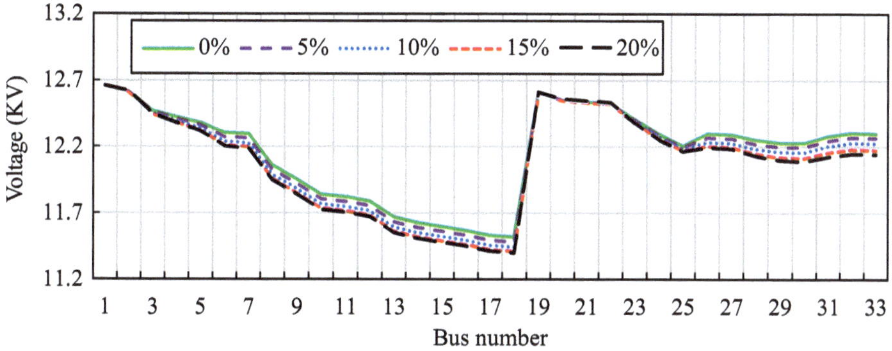

Fig. 4.12 Sensitivity analysis of the voltage profile in the EDN at $t = 22$

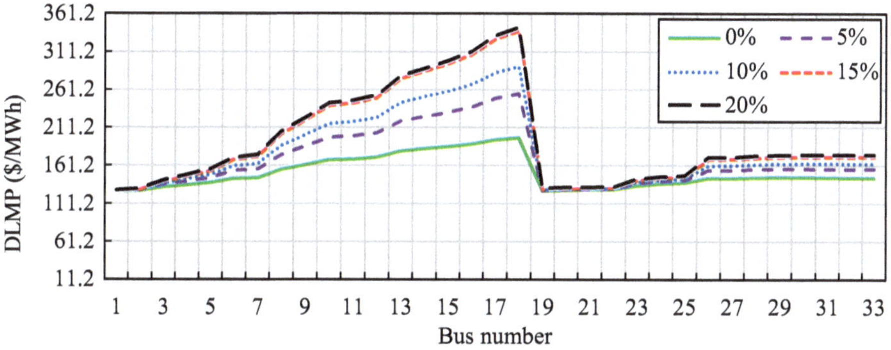

Fig. 4.13 Sensitivity analysis of the DLMP in the EDN at $t = 22$

In this case study, a sensitivity analysis is presented to investigate the reliance of the EDN on the NGDN. In the proposed sensitivity analysis, the behavior of the IUEAGDNs has been investigated for a step increase in residential natural gas consumption. Figure 4.12 illustrates the profile of the EDN buses voltage for the stepwise increase of residential loads of the gas network. The natural gas residential loads increase by 5% in each step. As residential consumption of natural gas increases in steps, power supply to industrial loads such as GFDGs is limited and causes a decrease in local production in the EDN. Therefore, the decrease in local production power leads to an increase in the electricity acquired from the upstream network, and as a result, the voltage drop also increases in the EDN nodes. In most of the local electrical markets (LEMs) in the world, the trading of electric power in EDNs is conducted through the nodal pricing method. Therefore, changes in the distribution local marginal price (DLMP) in the LEM lead to changes in the final cost of subscribers in the EDN.

Figure 4.13 shows the sensitivity analysis of the DLMP in the urban EDN at time $t = 22$. It's established that with the incremental change in residential gas consumption, the DLMP of EDN also increases. According to Fig. 4.14, the decrease in the

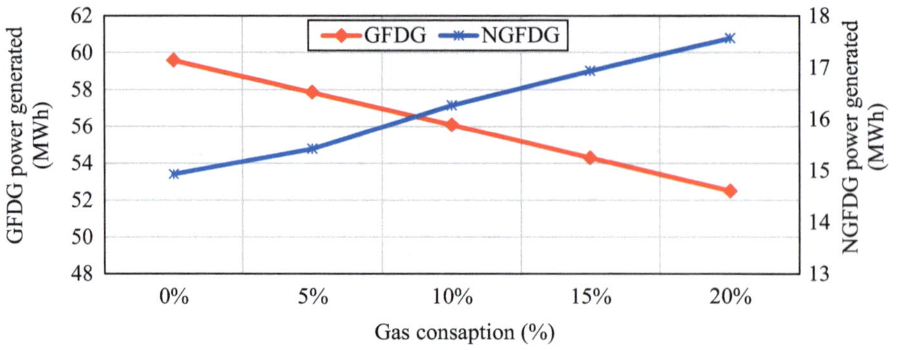

Fig. 4.14 Comparison of the total production capacity of GFDG and NGFDG

Table 4.2 Sensitivity analysis of operating cost of the IUEAGDS

	0%	5%	10%	15%	20%
Total operation cost ($/MWh)	13,886.83287	14,206.92659	14,535.09327	14,867.7274	15,203.98944
Purchased from WEM ($/MWh)	5069.188978	5235.542331	5337.887957	5476.095583	5623.895187
DG costs ($/MWh)	2775.158726	2869.120831	3035.559049	3165.275411	3287.895945
IL cost ($/MWh)	147.7738955	151.8059922	156.2035146	166.5815135	178.3045444
GS ($/kcf)	5894.671926	5950.416799	6005.401175	6059.732179	6113.849842

production capacity of cheap GFDGs leads to an incremental change in the production capacity of expensive producer, and as a result, the total price of electric energy for the customers will increase dramatically. Therefore, it can be concluded that changes in the residential consumption load of the gas network lead to DLMP fluctuations in the electricity market. Table 4.2 shows the sensitivity analysis of the operation cost of the IUEAGDNs. As observed, the total operation cost of the IUEAGDNs also increases with the gradual increase in natural gas network consumption loads.

4.6.2 Case Study 2

Case study 2 investigates the effect of using GSSs on the operation of the IUEAGDNs. Figure 4.15 compares the SOC of the GSS with the power consumption profile of the NGDN. It should be noted that in the GSS curve, the upward slope means charging and the downward slope means discharging the GSS. Therefore, the GSS stores gas fuel in off-peak times and discharges it in on-peak load times.

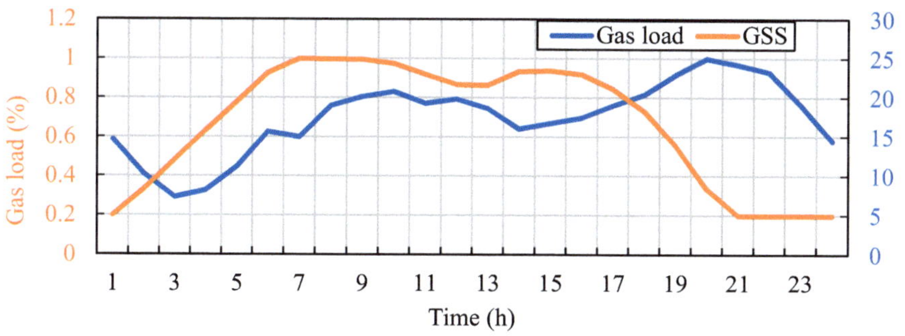

Fig. 4.15 Comparison of GSS storage with gas network consumption profile

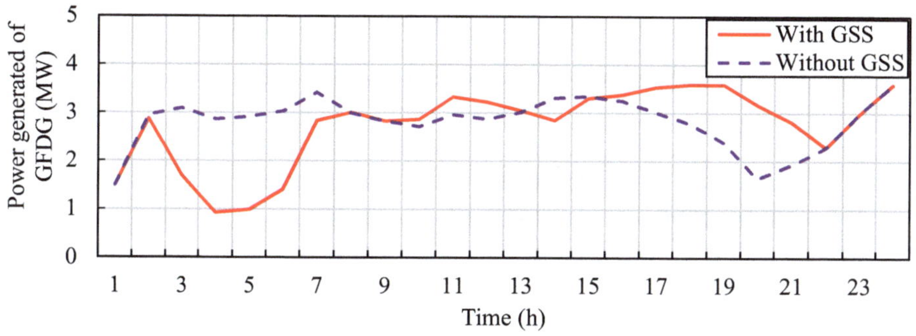

Fig. 4.16 Comparison of GFDG generation capacity with/without consideration of GSS

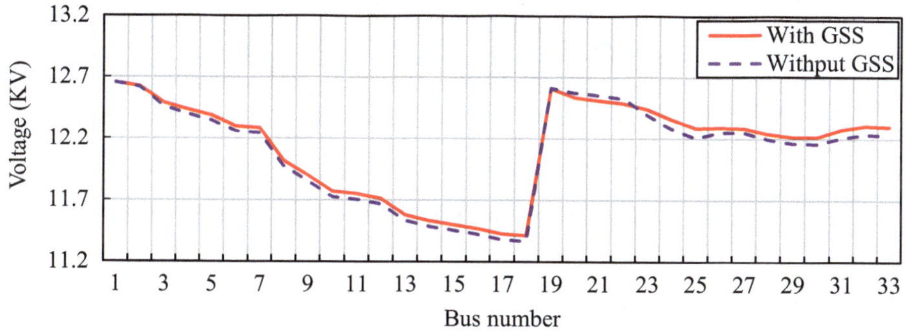

Fig. 4.17 Voltage profile of EDN with/without considering GSS

Figure 4.16 shows the comparison of the production power of GFDGs with/without GSS. As it is known, the presence of the GSS increases the fuel supply to GFDGs, especially during the on-peak hours of the EDN (16:00–22:00). Figure 4.17 indicates the profile of the EDN voltage with/without considering GSS. The presence of GSS increases fuel supply to GFDGs and increases local electricity

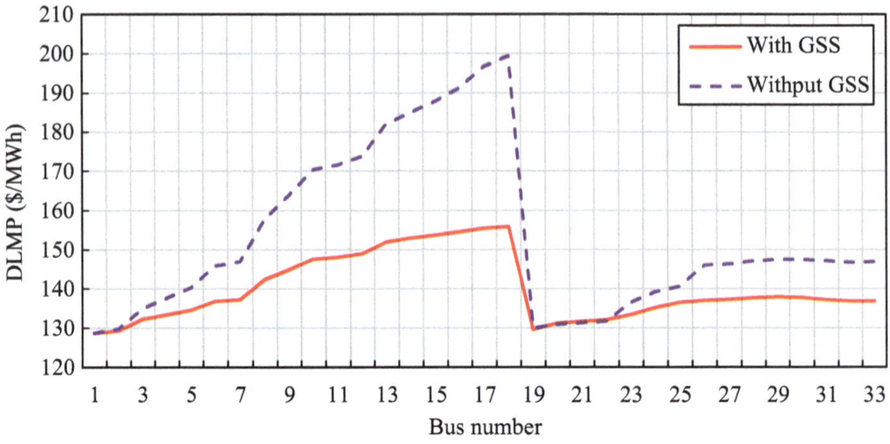

Fig. 4.18 DLMP of EDN with/without consideration of GSS

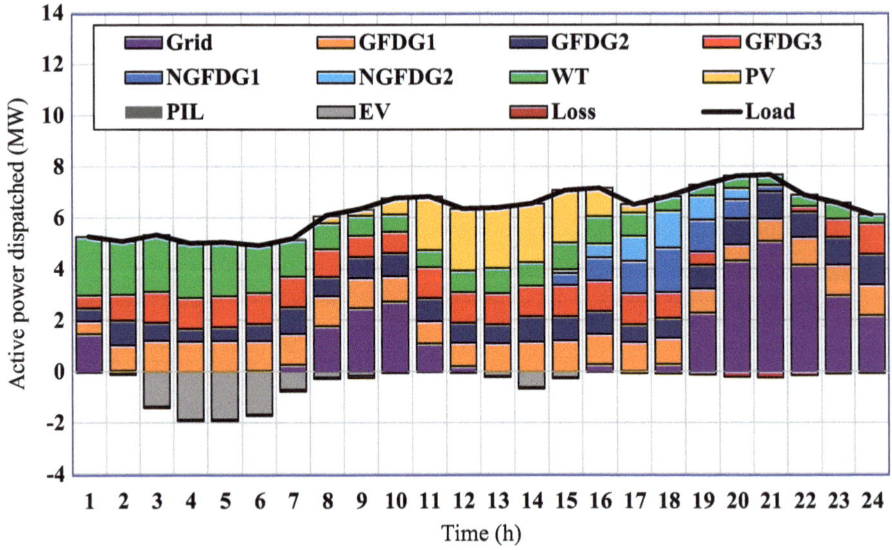

Fig. 4.19 Optimum scheduling of electrical load supply in the EDN

generation. As a result, with the increase in local production, the electricity acquired from the upstream network is reduced and the voltage profile is improved. Figure 4.18 illustrates the DLMP of the EDN with/without considering GSS. As it is known, the presence of GSS increases the production of cheap GFDGs, and as a result, the DLMP of the EDN also decreases in the presence of GSS.

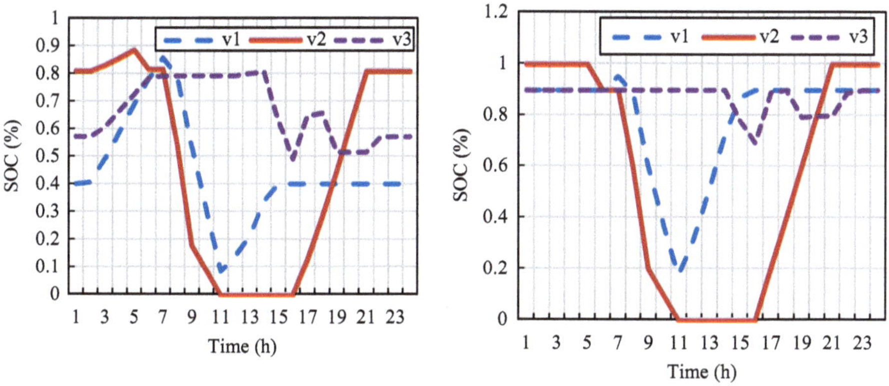

a) Smart charge strategy

b) Uncontrolled charge strategy

Fig. 4.20 EV battery charge under different strategies. (**a**) Smart charge strategy. (**b**) Uncontrolled charge strategy

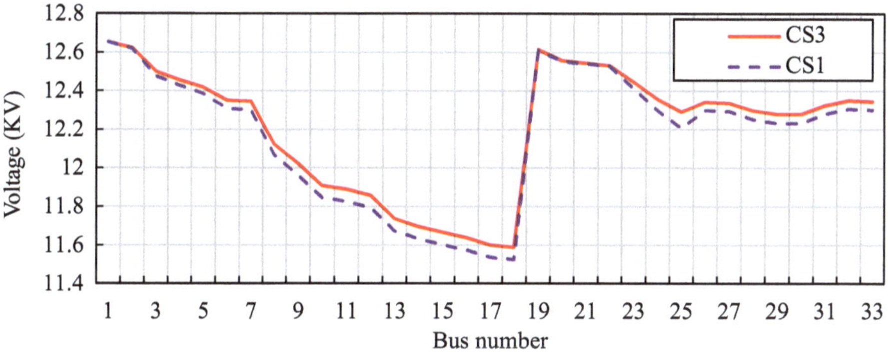

Fig. 4.21 Comparison of the voltage profile of the EDN in first and second case studies

4.6.3 Case Study 3

The case study 3 investigates stochastic optimal programming of IUEAGDNs in the presence of SPLs without regarding the smart charging option. As is known, with the execution of the smart charging strategy in the EDN, the off-peak time that has lower prices is allocated for charging EVs rather than peak and high-cost hours. The SOC of EVs with/without considering smart charging is compared in Fig. 4.19.

In the normal charging strategy, EVs start charging their batteries immediately after entering the parking lots. This way of charging creates a new peak hour, especially in the expensive hours of the EDN. In the smart charging strategy, EVs are charged when it's off-peak hours that has lower price. In the smart charging strategy, EVs undergo maximum charging during the early times of the night, i.e. 1–5 (please pay attention to Fig. 4.20b).

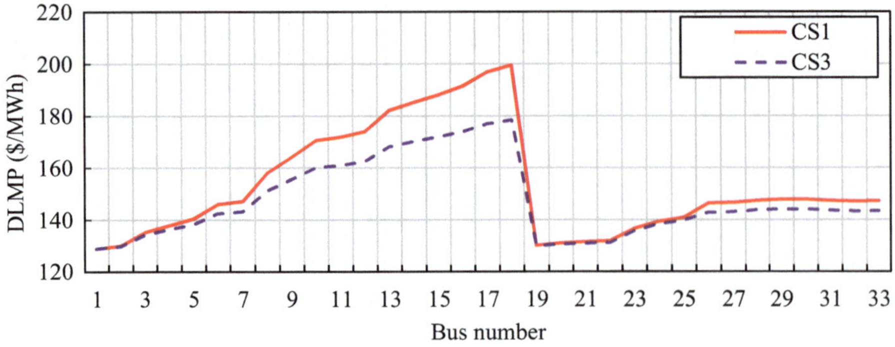

Fig. 4.22 Comparison of the DLMP in first and second case studies

Figure 4.21 indicates a comparison of the voltage profile of the EDN with first case study. As it is known, by implementing the smart charging strategy, the purchased electricity from the upstream network is reduced and electric loads are fed using local production. Hence, increasing the supply of EDN loads with DGs leads to enhancing the voltage profile within the third case of study. The comparison of the DLMP of EDN in case studies first and second is shown by Fig. 4.22. As it is known, by applying the smart charging strategy, the expensive power generation resource in the EDN has been reduced. Hence, the decrease in the power generated of expensive DGs leads to decrease in the DLMP of the EDN in contrast to the first case study. The total operation cost of the third case study is calculated to be \$11,737.59/MWh, which is a saving of \$2,183.39/MWh in contrast to the first case study.

4.6.4 Case Study 4

The fourth case study investigates a hybrid stochastic-robust optimization formwork for IUEAGDNs in the presence of SPL and GSS. Therefore, a strategy of robust optimization has been employed to represent the unpredictable nature of pricing in the electricity market, while the stochastic programming method has been utilized for modeling the unpredictable nature of renewable energy resources. Figure 4.23 illustrates the total operating costs with the total purchased electrical power of the EDN from the upstream network. In the robust optimization approach, when $\Gamma = 0$, it signifies that the effect of uncertainty in the proposed problem is

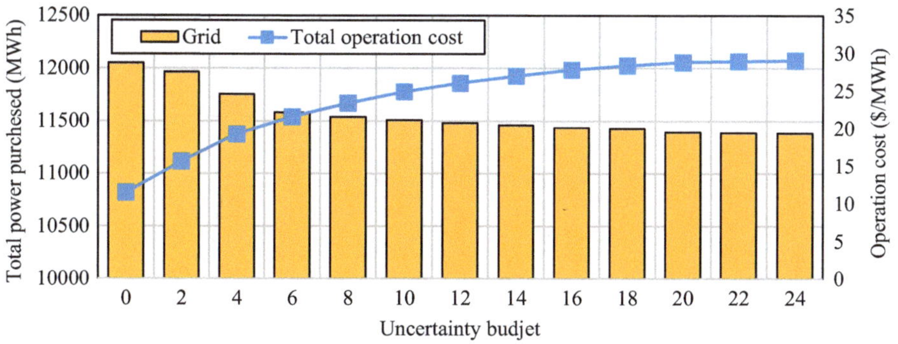

Fig. 4.23 Comparison of operating costs and purchasing power from the upstream network for different budgets

completely ignored. Conversely, when $\Gamma = T$, it means that the proposed approach is in the worst case. Figure 4.23 indicates that any incremental change of Γ leads to decreases in the electricity purchased from the upstream network, while the operational costs of the IUEAGDNs also increase. In other words, any increase in the integrated system operator's risk aversion leads to an increase in the operational costs of IUEAGDNs. While operating costs have indeed increased in the worst-case scenario, the optimal decision-making process of IUEAGDNs remains robust against the adverse effects of fluctuating prices in the wholesale electricity market. Table 4.3 shows the sensitivity analysis of total operating costs for different uncertainty budgets. As it is known, with the increase of Γ, the total operation cost in all stochastic scenarios also increases. According to Table 4.3, scenario S6 has the highest operational cost and scenario S7 has the lowest operational cost. Therefore, the worst and best scenarios for uncertain behavior of renewable energy sources are S6 and S7, respectively.

Table 4.4 shows a comparison of the operation costs of the IUEAGDNs in different case studies. As it is known, the integration of the GSS leads to a decrease of 6.72% of the total operation costs in contrast to the first case study. Additionally, with the execution of the smart charging strategy in the third case study, operation costs are reduced by 15.72% in contrast to the first case study, which is a very significant amount. Through the utilization of the GSS and implementing the smart charging strategy in case study 4-A, operating costs are reduced by 22.05% compared to the first case study. Finally, the total operating costs in the worst-case increase by 11.56% compared to case study 4-A.

Table 4.3 Sensitivity analysis of total operating costs for different uncertainty parameters

	S1	S2	S3	S4	S5	S6	S7	S8	S9	S10
$\Gamma = 0$	11,218.56	11,155.49	11,067.05	11,138.84	11,030.29	11,840.25	11,092.55	10,999.62	11,157	11,050.22
$\Gamma = 2$	11,424.62	11,355.09	11,292.35	11,345.84	11,252.34	12,052.67	11,291.92	11,219.48	11,381.63	11,288.48
$\Gamma = 4$	11,627.85	11,536.73	11,498	11,557.46	11,465.25	12,260.59	11,496.33	11,427.88	11,596.35	11,505.6
$\Gamma = 6$	11,713.22	11,609.5	11,584.89	11,644.19	11,547.33	12,336.25	11,577.66	11,537.55	11,683.48	11,597.07
$\Gamma = 8$	11,822.97	11,718.31	11,690.71	11,733.98	11,647.84	12,461.77	11,662.23	11,667.38	11,770.24	11,724.55
$\Gamma = 10$	11,910.27	11,818.82	11,779.25	11,808.83	11,729.63	12,583.07	11,734.07	11,755.56	11,830.97	11,811.7
$\Gamma = 12$	11,963.55	11,892.13	11,848.73	11,850.72	11,794.36	12,670.51	11,801.09	11,806.23	11,874.82	11,855.33
$\Gamma = 14$	12,020.35	11,954.93	11,911.71	11,912.94	11,843.7	12,737.76	11,853.87	11,869	11,927.33	11,906.05
$\Gamma = 16$	12,053.4	12,007.25	11,945.32	11,953.77	11,862.91	12,804.4	11,881.87	11,923.08	11,964.46	11,939.54
$\Gamma = 18$	12,089.04	12,045.23	11,985.93	11,988.72	11,889.59	12,849.1	11,914.92	11,962.35	11,998.34	11,973.01
$\Gamma = 20$	12,085.82	12,048.88	11,976.63	11,994.86	11,882.69	12,878.11	11,913.92	11,968.31	12,002.71	11,979.44
$\Gamma = 22$	12,086.42	12,051.74	11,976.9	11,996.93	11,884.5	12,885.48	11,914.53	11,969.95	12,004.04	12,010.5
$\Gamma = 24$	12,087.48	12,052.21	11,991.08	11,997.26	11,894.32	12,887.66	11,915.1	11,969.99	12,005.57	12,025.18

Table 4.4 Operational costs of the IUEAGDNs in different case studies

	CS1	CS2	CS3	CS4-a	CS4-b
Total operation cost ($/MWh)	13,886.83	12,952.9	11,703.43	10,824.39	12,076.48
Purchased from WEM ($/MWh)	5069.189	4250.913	3418.317	2828.842	1489.977
DG costs ($/MWh)	2775.159	2410.447	2073.603	1555.11	3259.901
IL cost ($/MWh)	147.7739	141.303	35.70888	35.25031	35.79263
GS ($/kcf)	5894.672	6150.204	6137.683	6369.316	6369.918
EV ($/MWh)	0	0	38.08649	35.84101	38.08391

4.7 Conclusion

In this study, a stochastic-robust hybrid programming based on convex optimization was modeled for IUEAGDNs in the company of gas storage system (GSS) and smart parking lot (SPL). In particular, the difference between the implementation of uncontrolled and controlled (smart) charging strategies in the IUEAGDNs was demonstrated. The SPL was presented using the k-means clustering method by considering the uncertainty characteristic of EV owners. Stochastic and robust optimization method was employed to simulate the unpredictable characteristics of renewable energy resources and the fluctuating costs in the wholesale electricity market. The effectiveness of the suggested framework was assessed through four case studies and summary of presented work is as follows:

1. The GSS system stores gas fuel in off-peak and cheap times and discharges the stored fuel in the NGDN during on-peak hours that have higher prices. This action leads to the reduction of the congestion of the gas network pipelines, especially during peak gas load hours, and increases fuel supply to industrial loads such as GFDGs. Additionally, the integration of GSS into the gas network cause to a 6.72% decrease in operation costs in the entire IUEAGDNs and increment the security of demand and energy supply.
2. The implementation of the smart charging strategy in SPL prevents the occurrence of a new peak load period in the electricity distribution network (EDN) and decreases the total operating costs by 15.72%. Moreover, the acquired results indicate that employing the smart charging strategy results in enhancing the voltage profile and decreasing the DLMP within the local market.
3. The total operation costs of the IUEAGDNs increase in the worst case of robust optimization. However, the optimal decision made by the system operator demonstrates robustness against the adverse effects of uncertainty parameters.

Appendix: Data and Information (Table 4.5)

Table 4.5 Data and parameters of UIEAGDN [35]

Parameter	Unit	Amount	Parameter	Unit	Magnitude
η_c	%	0.95	$C^{DG}_{g \in NGFU}$	$/MWh	180, 195
EB_e	kW	30	C^{IL}_{c}	$/MWh	350
SOC^{min}_{e}	%	0.2	C^{Gas}_{k}	$/kcf	2.5, 3
SOC^{max}_{e}	%	1	C^{ch}_{u}, C^{dis}_{u}	$/kcf	0.5
$ECPM_e$	kWh/mile	0.3	$\underline{P^{DG}_{g}}, \overline{P^{GD}_{g}} \in NGFDU$	MW	0, 2.8
SOC^{des}_{e}	%	0.9	$\underline{Q^{DG}_{g}}, \overline{Q^{GD}_{g}}$	Mvar	0, 1.4
Cr	kW	10	R^{UP}_{g}, R^{DN}_{g}	MM/h	0.5
BC	$/kWh	400	$\overline{S^{DG}_{g}}$	MVA	3
Γ	–	[0,24]	$\underline{P^{in}_{t}}, \overline{P^{in}_{t}}$	MW	−10, 10
$\underline{V_i}, \overline{V_i}$	KV	13.92, 11.36	$\underline{Q^{in}_{t}}, \overline{Q^{in}_{t}}$	MW	−5, 5
V_{base}	KV	12.66	$\eta^{ch}_{u}, \eta^{dis}_{u}$	%	0.9, 0.9
$G^{GSS}_{u,s,ini}$	kcf	5	$\underline{G^{GSS}_{u}}, \overline{G^{GSS}_{u}}$	kcf	0, 25
$\overline{S^{in}_{s,t}}$	MVA	12	$\overline{G^{ch}_{u}}, \overline{G^{dis}_{u}}$	kcf	5
ϕ	Kcf/MWh	3.41	η^{DG}_{g}	%	0.4

References

1. IEA-Natural Gas. https://www.iea.org/energy-system/fossil-fuels/natural-gas. Accessed on 6 Apr 2024
2. EIA-U.S. electricity generation.
3. M.J. Economides, D.A. Wood, The state of natural gas. J. Nat. Gas Sci. Eng. **1**(1-2), 1–13 (2009)
4. Statistical Review of World Energy 2021. 70th edition.
5. Reserves by country. https://www.worldometers.info/gas/gas-reserves-by-country. Accessed 6 Apr 2024
6. Y. He et al., Decentralized optimization of multi-area electricity-natural gas flows based on cone reformulation. IEEE Trans. Power Syst. **33**(4), 4531–4542 (2017)
7. C. Liu, M. Shahidehpour, Y. Fu, Z. Li, Security-constrained unit commitment with natural gas transmission constraints. IEEE Trans. Power Syst. **24**(3), 1523–1536 (2009)
8. IRENA. Innovation landscape brief: flexibility in conventional power plants, 2019
9. E. Raheli, Q. Wu, M. Zhang, C. Wen, Optimal coordinated operation of integrated natural gas and electric power systems: A review of modeling and solution methods. Renew. Sust. Energ. Rev. **145**, 111134 (2021)
10. W. Wei, J. Wang, *Modeling and optimization of interdependent energy infrastructures* (Springer, Cham, 2020)

11. T. Li, M. Eremia, M. Shahidehpour, Interdependency of natural gas network and power system security. IEEE Trans. Power Syst. **23**(4), 1817–1824 (2008)
12. M. Shahidehpour, Y. Fu, T. Wiedman, Impact of natural gas infrastructure on electric power systems. Proc. IEEE **93**(5), 1042–1056 (2005)
13. S.A. Mansouri, A. Ahmarinejad, E. Nematbakhsh, M.S. Javadi, A.E. Nezhad, J.P. Catalão, A sustainable framework for multi-microgrids energy management in automated distribution network by considering smart homes and high penetration of renewable energy resources. Energy **245**, 123228 (2022)
14. J. Caballero-Pena, C. Cadena-Zarate, A. Parrado-Duque, G. Osma-Pinto, Distributed energy resources on distribution networks: A systematic review of modelling, simulation, metrics, and impacts. Int. J. Electr. Power Energy Syst. **138**, 107900 (2022)
15. P. Harsh, D. Das, Energy management in microgrid using incentive-based demand response and reconfigured network considering uncertainties in renewable energy sources. Sustain Energy Technol Assess **46**, 101225 (2021)
16. B. Javanmard, M. Tabrizian, M. Ansarian, A. Ahmarinejad, Energy management of multi-microgrids based on game theory approach in the presence of demand response programs, energy storage systems and renewable energy resources. J. Energy Storage **42**, 102971 (2021)
17. M.M. Hayati, A. Aminlou, K. Zare, M. Abapour, A Two-Stage Stochastic Optimization Scheduling Approach for Integrating Renewable Energy Sources and Deferrable Demand in the Spinning Reserve Market, in *2023 8th International Conference on Technology and Energy Management (ICTEM)*, (IEEE, Piscataway, 2023), pp. 1–7
18. C.D. Iweh, S. Gyamfi, E. Tanyi, E. Effah-Donyina, Distributed generation and renewable energy integration into the grid: Prerequisites, push factors, practical options, issues and merits. Energies **14**(17), 5375 (2021)
19. Z. Li, S. Su, X. Jin, H. Chen, Distributed energy management for active distribution network considering aggregated office buildings. Renew. Energy **180**, 1073–1087 (2021)
20. N. Patari, V. Venkataramanan, A. Srivastava, D.K. Molzahn, N. Li, A. Annaswamy, Distributed optimization in distribution systems: Use cases, limitations, and research needs. IEEE Trans. Power Syst. **37**(5), 3469–3481 (2021)
21. Y. Li, X. Tang, X. Lin, L. Grzesiak, X. Hu, The role and application of convex modeling and optimization in electrified vehicles. Renew. Sust. Energ. Rev. **153**, 111796 (2022)
22. B. Singh, P.K. Dubey, Distributed power generation planning for distribution networks using electric vehicles: Systematic attention to challenges and opportunities. J. Energy Storage **48**, 104030 (2022)
23. P.H. Hoang et al., A dual distributed optimal energy management method for distribution grids with electric vehicles. IEEE Trans. Intell. Transp. Syst. **23**(8), 13666–13677 (2021)
24. S. Claeys, M. Vanin, F. Geth, G. Deconinck, Applications of optimization models for electricity distribution networks. Wiley Interdiscip. Rev. Energy Environ. **10**(5), e401 (2021)
25. M.T. Hussain, N.B. Sulaiman, M.S. Hussain, M. Jabir, Optimal management strategies to solve issues of grid having electric vehicles (EV): A review. J. Energy Storage **33**, 102114 (2021)
26. Z. Zhang, C. Wang, H. Lv, F. Liu, H. Sheng, M. Yang, Day-ahead optimal dispatch for integrated energy system considering power-to-gas and dynamic pipeline networks. IEEE Trans. Ind. Appl. **57**(4), 3317–3328 (2021)
27. J. Duan, Y. Yang, F. Liu, Distributed optimization of integrated electricity-natural gas distribution networks considering wind power uncertainties. Int. J. Electr. Power Energy Syst. **135**, 107460 (2022)
28. V. Shabazbegian, H. Ameli, M.T. Ameli, G. Strbac, Stochastic optimization model for coordinated operation of natural gas and electricity networks. Comput. Chem. Eng. **142**, 107060 (2020)
29. M. Sattar, M.S. Moghaddam, A. Azarfar, N. Salehi, M. Vahedi, Co-optimization of integrated energy systems in the presence of renewable energy, electric vehicles, power-to-gas systems and energy storage systems with demand-side management. Clean Energy **7**(2), 426–435 (2023)

30. N. Nasiri, S. Zeynali, S.N. Ravadanegh, S. Kubler, Y. Le Traon, A convex multi-objective distributionally robust optimization for embedded electricity and natural gas distribution networks under smart electric vehicle fleets. J. Clean. Prod. **434**, 139843 (2024)
31. N. Nasiri, S. Zeynali, S.N. Ravadanegh, S. Kubler, Moment-based distributionally robust peer-to-peer transactive energy trading framework between networked microgrids, smart parking lots and electricity distribution network. IEEE Trans. Smart Grid **15**(2), 1965–1977 (2023)
32. N. Nasiri, S. Zeynali, S.N. Ravadanegh, S. Kubler, Economic-environmental convex network-constrained decision-making for integrated multi-energy distribution systems under electrified transportation fleets. J. Clean. Prod. **379**, 134582 (2022)
33. N. Nasiri et al., A robust bi-level optimization framework for participation of multi-energy service providers in integrated power and natural gas markets. Appl. Energy **340**, 121047 (2023)
34. N. Nasiri, S. Zeynali, S.N. Ravadanegh, N. Rostami, A robust decision framework for strategic behaviour of integrated energy service provider with embedded natural gas and power systems in day-ahead wholesale market. IET Gener. Transm. Distrib. **16**(3), 561–579 (2022)
35. N. Nasiri, S. Zeynali, S.N. Ravadanegh, Unit commitment scheduling in embedded natural gas and electricity distribution systems considering smart plug-in electric vehicle fleets and linepack storage. Sustain. Energy Technol. Assess. **68**, 103875 (2024)

Chapter 5
Multi-Energy Networked Microgrid Scheduling in Smart Cities Under Transactive Energy Concept

Nima Nasiri, Asma Nasiri, Sajad Najafi Ravadanegh, Behnam Mohammadi-ivatloo, and Mehdi Abapour

Abbreviations Sets and Indexes

i, j, k	Indices of electrical microgrid bus
t, l	Indices of time, line
m, m'	Indices of microgrids
g, r, d	Indices of DG, PV, electrical demand
b, e	Indices of gas boiler, electrical boiler
R	Index of regional CHP
$\Omega_{\text{CHP}}, \Omega_{\text{NCHP}}$	Set of CHP and non CHP generators

N. Nasiri (✉) · S. N. Ravadanegh
Resilient Smart Grids Research Lab, Electrical Engineering Department, Azarbaijan Shahid Madani University, Tabriz, Iran
e-mail: s.najafi@azaruniv.ac.ir

A. Nasiri · M. Abapour
Faculty of Electrical and Computer Engineering, Energy System Research Institute (ESRI), University of Tabriz, Tabriz, Iran
e-mail: asmanasiri1402@ms.tabrizu.ac.ir; abapour@tabrizu.ac.ir

B. Mohammadi-ivatloo
Faculty of Electrical and Computer Engineering, Energy System Research Institute (ESRI), University of Tabriz, Tabriz, Iran

School of Energy Systems, Department of Electrical Engineering, LUT University of Technology, Lappeenranta, South Karelina, Finland
e-mail: behnam.ivatloo@lut.fi

A. Ahmadian et al. (eds.), *Sustainable Energy Resources in Smart Cities*, Green Energy and Technology, https://doi.org/10.1007/978-3-031-91516-1_5

Parameters

$r^{l}_{m,i,j}, x^{l}_{m,i,j}$	Magnitude of resistance and inductance of MEMNG
$z^{l}_{m,i,j}$	Magnitude of impedance of MEMNG
$\overline{P^{DG}_{g}}, \underline{P^{DG}_{g}}$	Maximum and minimum active power generation amount of DGs (MW)
$\overline{Q^{DG}_{g}}, \underline{Q^{DG}_{g}}$	Maximum and minimum reactive power generation amount of DGs (MVar)
P^{R}_{m}	Power generation corresponding to the feasible operating region of CHP unit
H^{R}_{m}	Heat generation corresponding to the feasible operating region of CHP unit
$\eta^{GB}_{m,b}$	Gas boiler efficacy (%)
$\eta^{e}_{m,e}$	Electrical boiler efficacy (%)
$\gamma^{P}_{g,m}, \gamma^{H}_{g,m}$	Power and heat coefficient of CHP
$\gamma^{B}_{b,m}$	Gas boiler coefficient
$R^{UP}_{m,g}, R^{DN}_{m,g}$	Ramp-up and ramp-down amounts of DGs (MW)
$\overline{P^{in}_{t}}, \underline{P^{in}_{t}}$	Maximum and minimum active power trade-off between END and upstream network (MW)
$\overline{Q^{in}_{t}}, \underline{Q^{in}_{t}}$	Maximum and minimum reactive power trading between END and upstream network (MVar)
$\overline{P^{PV}_{m,r}}$	Maximum power generation amounts of PV (MMW)
$P^{d}_{m,i,j}, Q^{d}_{m,i,t}$	Active and reactive power demand data (MW)
$Q^{d}_{m,i,j}$	Reactive power demand (Mvar)
$H^{load}_{m,t}$	Thermal demand (MW)
$C^{in}_{m,t}$	Cost of purchased electricity from upstream network ($/MWh)
$C^{Gas}_{m,t}$	Cost of received gas for CHP ($/MWh)
$C^{DG}_{m,g,t}$	Power generated cost by DG ($/MWh)
$C^{P2P}_{m,m',t}$	Cost of bilateral contract peer to peer ($/MWh)
$\overline{P^{P2P}_{m',m}}, -\overline{P^{P2P}_{m',m}}$	Maximum and minimum amount of P2P transactive energy (MW)

Variables

$P^{\mathrm{l}}_{\mathrm{m,i,j,t}}, Q^{\mathrm{l}}_{\mathrm{m,i,j,t}}$	Reactive and active power flow (MW)
$I^{\mathrm{l}}_{\mathrm{m,i,j,s,t}}$	Electricity current flow between two buses i, j in MEMNG (A)
$v_{m,\,i,\,j,\,t}$	Buses voltage of MEMNG (V)
$P^{DG}_{g,t}, Q^{DG}_{g,t}$	Active and reactive output power generated by DGs (MW)
$\alpha^{\mathrm{R}}_{\mathrm{m,t}}$	Operation point of CHP unit
$G^{\mathrm{GB}}_{\mathrm{m,b,t}}$	Gas consumption of gas boiler (KCF)
$H^{\mathrm{GB}}_{\mathrm{m,b,t}}$	Heat generated with gas boiler (MW)
$P^{\mathrm{EB}}_{\mathrm{m,e,t}}$	Power consumption of electrical boiler (MW)
$H^{\mathrm{EB}}_{\mathrm{m,e,t}}$	Heat generated with electrical boiler (MW)
$H^{\mathrm{DG}}_{\mathrm{m,g,t}}$	Heat generated by DGs (MW)
$G^{\mathrm{Gas}}_{\mathrm{m,t}}$	Gas consumption of ME (KCL)
P^{in}_{t}, Q^{in}_{t}	Active and reactive power trade-off between MEMNG and upstream network (MW-Mvar)
$P^{PV}_{m,r,t}$	Output power of PV (MW)
$P^{d}_{m,i,j}, Q^{\mathrm{d}}_{\mathrm{m,i,t}}$	Total active and reactive power demand (MW)
$P^{\mathrm{P2P}}_{\mathrm{m,m',t}}$	Amount of transactive power between MEMNG (MW)

5.1 Introduction

Conventionally, the power grid consisted of a generation part that was centralized, and so far away from consumers, also it involved transmission and distribution parts which were made for only transferring electricity. The centralized generation aspects of conventional power systems lead to a one-directional power flow, and due to that, we can easily operate the system [1]. However, the conventional grid's energy source consumption leads to environmental pollution and health issues [2]. Additionally, centralized generation can cause reliability issues, so we need a modern reliable power grid with decentralized sources that integrate renewable energies. Microgrids are one of the new concepts in modern power systems.

A microgrid is a small-scale electricity distribution system consisting of load, distributed generation resources (DERs) such as wind turbines, photovoltaic (PV), diesel generators (DGs), etc. Microgrids can meet required power demands in both isolated and grid-connected states. The implementation of microgrids can enhance grid efficiency, assist the system in emergencies, and help to meet the growing

Fig. 5.1 Illustration of a microgrid diagram

demands for electricity energy [3]. Additionally, microgrids can cause a reduction in congestion and losses of power systems due to the vicinity of power suppliers to the consumers [4].

The equipment of microgrids including types of energy suppliers and loads can vary in a broad range [5]. Figure 5.1 illustrates a sample of microgrids that consist of various components comprising PV, wind turbines, DGs, energy storage, electric vehicles (EVs), EV parking lots, etc. This microgrid is interconnected with the upstream network to purchase or inject electricity into the power grid. It is worth noting that the microgrids can also connect to other microgrids for selling or purchasing electricity.

Nowadays, power systems are confronting incremental growth in using small-scale DERs, and most companies and consumers are moving toward self-generating electricity by using DERs. This issue can help to reduce the power grid dependency and could be a significant step to smartening the cities. Additionally, from the perspective of the grid, DERs can be very useful in providing flexibility and helping the system with energy management issues [6]. In recent years, the role of consumers has changed, and the consumers are turning into prosumers [7]. The prosumers can both generate and consume electricity, whereas consumers solely consume it. The prosumers can trade electricity to another prosumer or sell it to the consumer. The direct trading of electricity energy between prosumers and consumers is referred peer-to-peer (P2P) transactive energy.

P2P transactive energy trading comprises different participation, encompassing electricity generators, prosumers, and consumers. Within this framework, peers directly purchase or sell electricity without any intermediary or reliance on the power grid [8]. The P2P concept usually is applied in local power systems and it's a case of shared economy [7]. In the conventional way of energy trading, the transmission and distribution systems deliver or sell the electricity to consumers and the

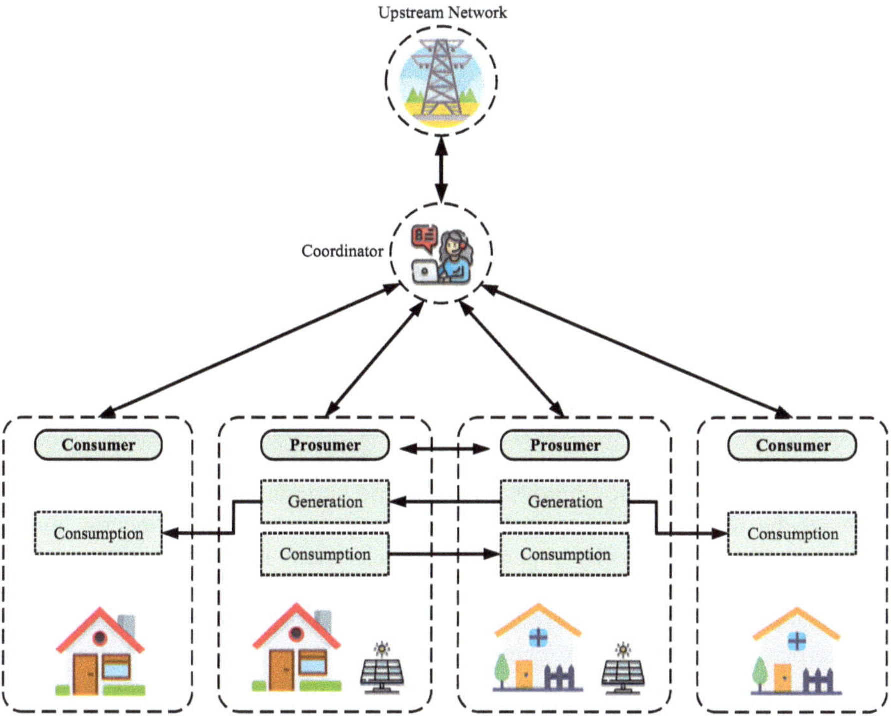

Fig. 5.2 An illustration of P2P concept

consumers can only purchase energy from the grid. On the other hand, P2P energy trading is a multidirectional way of energy sharing that allows consumers to participate in energy trading as prosumers. Figure 5.2 explains the concept of P2P and shows a group of consumers and prosumers which are trading energy with each other. This group of consumers and prosumers trade energy with each other and also with the upstream network through the distribution and transition systems.

5.2 Literature Review

A comprehensive review of various optimization methods for modeling the microgrid is presented in [9] by focusing on energy management challenges, load forecasting, demand side management, economic dispatch, and unit commitment. A stochastic approach for optimizing the energy management problem of the multi-microgrid is proposed in [10] for minimizing the operational cost and carbon emission of the suggested framework. An optimum energy management technique of the multi-microgrids is presented in [11] by considering the cooperative game theory concept. This paper optimized the total cost of the system and maximized the benefits of microgrids. A two-stage optimum energy management scheduling of microgrids is suggested in [12] in the presence of energy markets. The authors in [12] modeled the mentioned process with a bi-level optimization framework consisting of day-ahead optimal planning of multi-microgrids by using game theory and managing the unpredictable behavior of electricity demand and renewable energies.

In [13], a scheduling of multi-energy microgrids (MMGs) is presented with considering the plug-in hybrid vehicles and renewable energies under the uncertainties. The authors of this paper used the game theory concept for optimizing the total cost. In [14], an optimal resilient approach is proposed for optimizing the operation of multi-energy microgrids (MMGs) with DERs to manage and meet thermal and electrical demands. An optimum energy management technique is presented in [15] to minimize and optimize the operational cost of an MMG system by taking into consideration the carbon emissions and operational constraints. An energy planning framework is proposed in [16] for the MMG system that integrates electric, heat, cooling, and water parts to improve the microgrid system.

A stochastic approach is stated in [17] for planning of multi-energy multi-microgrid that coordinates hydrogen, heat, and electricity in a standalone system. In [18], a framework is suggested for minimizing the operational cost of MMG, including integrated electrical and thermal systems by considering network constraints. A peer-to-peer (P2P) transactive energy scheduling is proposed in [19] for multi-microgrids considering network constrain and uncertainty. A distributed day-ahead energy trading approach is presented in [20] for multi-microgrid management in a distribution system. An energy trading framework of microgrids with P2P transactive energy is suggested in [21] by considering EVs, demand response programs, and emergency energy sources. A day-ahead energy management approach of a microgrid is suggested in [22] considering demand response, uncertainty, and use of fast alternative direction method of multipliers (F-ADMM) algorithm. In [23], a decentralized robust approach is stated for coordinating active distribution networks with multi-microgrids.

In studies [9–18], the authors focus on trading microgrids with different electricity markets, considering physical and economic constraints. However, the advantages of peer-to-peer trading between microgrids in transactive energy markets (TEMs) have been ignored. Participating in transactive energy markets enables microgrids to effectively utilize their surplus generation capacity while upholding privacy and maximizing profits.

In researches [9–12, 19–22], the authors have examined different models of microgrids for technical and economic studies. However, they only focused on one type of energy carrier, electricity. Combined heat and power (CHP) units within microgrids not only fulfill electric loads but also support heating demands, enhancing overall energy efficiency and resilience.

Some authors have conducted various studies on the physical, economic, and technical structures of microgrids in researches [9–16, 21, 22]. However, they have ignored network constraints such as voltage, current, and flow of active and reactive

Table 5.1 Comparing this research with previous studies

References	Microgrid	Multi-energy	CHP	Network constrains	P2P transactive energy	Fast ADMM algorithm
[9]	✓	×	×	×	×	×
[10]	✓	×	×	×	×	×
[11]	✓	×	×	×	×	×
[12]	✓	×	×	×	×	×
[13]	✓	✓	×	×	×	×
[14]	✓	✓	✓	×	×	×
[15]	✓	✓	✓	×	×	×
[16]	✓	✓	✓	×	×	×
[17]	✓	✓	✓	✓	×	×
[18]	✓	✓	✓	✓	×	×
[19]	✓	×	×	✓	✓	×
[20]	✓	×	×	✓	✓	×
[21]	✓	×	×	×	✓	✓
[22]	✓	×	×	×	✓	✓
[23]	✓	×	×	×	✓	✓

power. Microgrids are small-scale distribution networks, and ignoring network constraints can have adverse effects on research results.

According to the above gaps and Table 5.1, the contribution of this research is as follows:

A decentralized optimization framework is proposed for P2P trading of multi-energy networked microgrids (MENMGs) in transactive energy market (TEM) and local electricity market (LEM) considering PV generations.

A distributed fast alternative direction method of multipliers (F-ADMM) algorithm is presented for fast convergence and achieving strong equilibrium point in solving the decentralized optimization problem.

5.3 Mathematical Modeling

5.3.1 Constraints of the Multi-Energy Networked Microgrid (MENMG)

Generally, microgrids are small-scale electrical distribution systems that generate and consume energy at low voltage levels. In this study, it is assumed that six microgrids with an m index exchange electric energy in a peer-to-peer (P2P) framework. Each of the microgrid consists of a set of buses $NB = \{0, 1, \ldots, N\}$ and a set of feeders$l = \{1, .., L\}$. In microgrids, bus 0 is the connection point of the distribution network to the microgrid, and its voltage is represented by 1 p.u. Therefore, for each bus i, the voltage value is indicated by v_i, while P_i and Q_i also indicate the active and reactive power injection in microgrids. Additionally, the active and reactive powers consumed for each bus of microgrids are also expressed using P_d and Q_d. Moreover, X_{ij} and R_{ij} express the resistance and reactance of microgrid lines, respectively, while P_{ij} and Q_{ij} model the flow of active and reactive power in microgrid feeders. The AC current flow through microgrid feeders is also modeled using I. It should be noted that in this study, the Distflow method was used to model the limitation of microgrid networks, which is expressed by limitations (5.1)–(5.4) [24]. The proof of Distflow method is presented in Appendix A. The active and reactive power balance equations in the microgrid are modeled by constraints (5.1) and (5.2), respectively.

$$P^{\mathrm{l}}_{\mathrm{m,i,j,t}} + P^{\mathrm{g}}_{\mathrm{m,i,t}} = P^{\mathrm{d}}_{\mathrm{m,i,t}} + \sum_{\mathrm{k}\in \mathrm{A}_i^k} \left(P^{\mathrm{l}}_{\mathrm{m,i,k,t}} + r^{\mathrm{l}}_{\mathrm{m,i,j}} I^{\mathrm{l}}_{\mathrm{m,i,j,s,t}}\right) \quad \forall m, \forall l, \forall t \tag{5.1}$$

$$Q^{\mathrm{l}}_{\mathrm{m,i,j,t}} + Q^{\mathrm{g}}_{\mathrm{m,i,t}} = Q^{\mathrm{d}}_{\mathrm{m,i,t}} + \sum_{\mathrm{k}\in \mathrm{A}_i^k} \left(Q^{\mathrm{l}}_{\mathrm{m,i,k,t}} + x^{\mathrm{l}}_{\mathrm{m,i,j}} I^{\mathrm{l}}_{\mathrm{m,i,j,s,t}}\right) \quad \forall m, \forall l, \forall t \tag{5.2}$$

$$v^2_{\mathrm{m,j,t}} = v^2_{\mathrm{m,i,t}} - 2\left(P^{\mathrm{l}}_{\mathrm{m,i,j,t}} r^{\mathrm{l}}_{\mathrm{m,i,j}} + Q^{\mathrm{l}}_{\mathrm{m,i,j,t}} x^{\mathrm{l}}_{\mathrm{m,i,j}}\right) + \left(z^{\mathrm{l}}_{\mathrm{m,i,j}}\right)^2 I^{\mathrm{l}}_{\mathrm{m,i,j,t}} \quad \forall m, \forall l, \forall t \tag{5.3}$$

$$I^{\mathrm{l}}_{\mathrm{m,i,j,t}} = \left(P^{\mathrm{l}}_{\mathrm{m,i,j,t}}\right)^2 + \left(Q^{\mathrm{l}}_{\mathrm{m,i,j,t}}\right)^2 / v^2_{\mathrm{m,i,j,t}} \quad \forall m, \forall l, \forall t \tag{5.4}$$

The constraint (5.4) involves the flow equation $I^{l}_{m,i,j,t}$ that renders the non-convex problem to a convex problem. It should be noted that active and reactive power losses in microgrids can be ignored because power losses are very low due to the small-scale power exchanges in microgrids. Hence, an approximation of a voltage profile, for example, $v_i = 1$ can be found to be $V_i^2 - V_j^2 = 2(V_i - V_j)$. With this assumption incorporated into the model described above, the linearized Distflow model can be reformulated below.

$$P^{\mathrm{l}}_{\mathrm{m,i,j,t}} + P^{\mathrm{g}}_{\mathrm{m,i,t}} = P^{\mathrm{d}}_{\mathrm{m,i,t}} + \sum_{\mathrm{k}\in \mathrm{A}_i^k} P^{\mathrm{l}}_{\mathrm{m,i,k,t}} \quad \forall m, \forall l, \forall t \tag{5.5}$$

$$Q^{\mathrm{l}}_{\mathrm{m,i,j,t}} + Q^{\mathrm{g}}_{\mathrm{m,i,t}} = Q^{\mathrm{d}}_{\mathrm{m,i,t}} + \sum_{\mathrm{k}\in \mathrm{A}_i^k} Q^{\mathrm{l}}_{\mathrm{m,i,k,t}} \quad \forall m, \forall l, \forall t \tag{5.6}$$

$$v_{\mathrm{m,i,t}} - v_{\mathrm{m,j,t}} = P^{\mathrm{l}}_{\mathrm{m,i,j,t}} r^{\mathrm{l}}_{\mathrm{m,i,j}} + Q^{\mathrm{l}}_{\mathrm{m,i,j,t}} x^{\mathrm{l}}_{\mathrm{m,i,j}} \quad \forall m, \forall l, \forall t \tag{5.7}$$

5.3.2 Constraints of Distributed Generation (DG) in MENMGs

Generally, the constraints of DGs in MENMGs are modeled by Eqs. (5.8)–(5.23). Respectively, Eqs. (5.8) and (5.9) model the constraints of active and reactive production power of DGs in MENMGs. It is worth noting that DGs in this microgrid are modeled by a microturbine (MT) and a CHP. In order to model the behavior of CHP electrical and heating power, the feasible operation region (FOR) method has

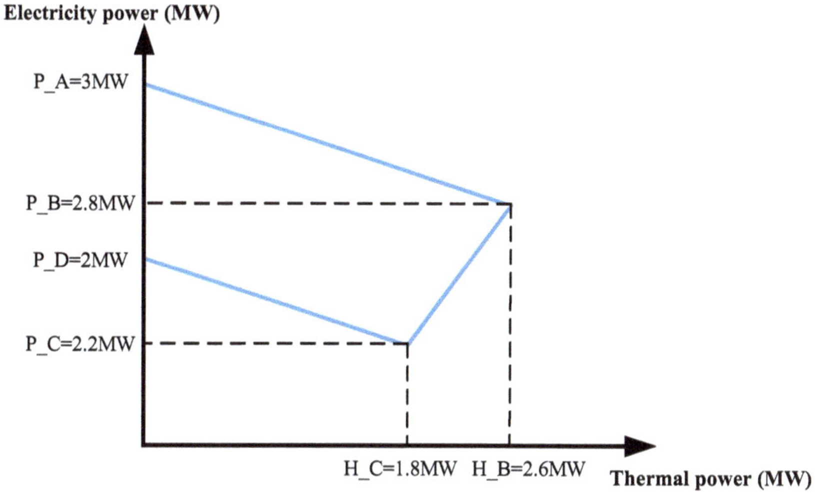

Fig. 5.3 Feasible operation region of CHP

been used. Figure 5.3 shows the operating constraints of CHP, which are mathematically expressed by Eqs. (5.10)–(5.13). Where Eqs. (5.10) and (5.11) show the electrical and heating production of CHP using the FOR method, respectively. Equations (5.11) and (5.12) also set the operating points of CHP on the FOR curve. The heating power produced by gas and electric boilers is also modeled using constraints (5.14) and (5.15). Equation (5.16) models the total gas consumption of MENMGs. Ramp rate limits of DGs are expressed using Eqs. (5.17) and (5.18). The constraint of active and reactive powers purchased from the local electricity market is also modeled by Eqs. (5.19) and (5.20). Constraint (5.21) models the maximum production power of the PV system. The total active and reactive power injected into the MENMG is modeled by (5.22) and (5.23), and also Eqs. (5.24) and (5.25) model the total active and reactive power consumption of the MENMG. The heating power balance equation in microgrids is also modeled using Eq. (5.26).

$$0 \le P_{g,t}^{DG} \le \overline{P_{g}^{DG}} \quad \forall g \in \Omega_{NCHP}, \quad \forall s, \forall t \tag{5.8}$$

$$0 \le Q_{g,t}^{DG} \le \overline{Q_{g}^{DG}} \quad \forall g, \forall s, \forall t \tag{5.9}$$

$$P_{m,g,t}^{DG} = \sum_{R} \alpha_{m,t}^{R} P_{m}^{R} \quad \forall g \in \Omega_{CHP}, \forall m, \forall t \tag{5.10}$$

$$H_{m,g,t}^{DG} = \sum_{R} \alpha_{m,t}^{R} H_{m}^{R} \quad \forall g \in \Omega_{CHP}, \forall m, \forall t \tag{5.11}$$

$$\sum_{R} \alpha_{m,t}^{R} = 1 \quad \forall u \in \Omega_{CHP}, \forall m, \forall t \tag{5.12}$$

$$0 \le \alpha_{m,t}^{R} \le 1 \quad \forall u \in \Omega_{CHP}, \forall R, \forall m, \forall t \tag{5.13}$$

$$H^{\mathrm{GB}}_{\mathrm{m,b,t}} = \eta^{\mathrm{GB}}_{\mathrm{m,b}} G^{\mathrm{GB}}_{\mathrm{m,b,t}} \quad \forall \mathrm{m}, \forall \mathrm{b}, \forall t \tag{5.14}$$

$$H^{\mathrm{EB}}_{\mathrm{m,e,t}} = \eta^{\mathrm{e}}_{\mathrm{m,e}} P^{\mathrm{EB}}_{\mathrm{m,e,t}} \quad \forall \mathrm{m}, \forall \mathrm{e}, \forall t \tag{5.15}$$

$$G^{\mathrm{Gas}}_{\mathrm{m,t}} = \sum_{g} \gamma^{P}_{\mathrm{g,m}} P^{\mathrm{DG}}_{\mathrm{m,g,t}} + \gamma^{H}_{\mathrm{g,m}} H^{\mathrm{DG}}_{\mathrm{m,g,t}} + \sum_{\mathrm{b}} \gamma^{B}_{\mathrm{b,m}} G^{\mathrm{GB}}_{\mathrm{m,b,t}} \quad \forall \mathrm{m}, \forall t \tag{5.16}$$

$$P^{\mathrm{DG}}_{\mathrm{m,g,t}} - P^{\mathrm{DG}}_{\mathrm{m,g,t\text{-}1}} \le R^{\mathrm{UP}}_{\mathrm{m,g}} \quad \forall \mathrm{m}, \forall g, \forall t \tag{5.17}$$

$$P^{\mathrm{DG}}_{\mathrm{m,g,t\text{-}1}} - P^{\mathrm{DG}}_{\mathrm{m,g,t}} \le R^{\mathrm{DN}}_{\mathrm{m,g}} \quad \forall \mathrm{m}, \forall \mathrm{u}, \forall t \tag{5.18}$$

$$-\overline{P^{\mathrm{in}}_{\mathrm{t}}} \le P^{\mathrm{in}}_{\mathrm{m,t}} \le \overline{P^{\mathrm{in}}_{\mathrm{t}}} \quad \forall \mathrm{m}, \forall t \tag{5.19}$$

$$\underline{Q^{\mathrm{in}}_{\mathrm{t}}} \le Q^{\mathrm{in}}_{\mathrm{m,t}} \le \overline{Q^{\mathrm{in}}_{\mathrm{t}}} \quad \forall \mathrm{m}, \forall t \tag{5.20}$$

$$0 \le P^{\mathrm{PV}}_{\mathrm{m,r,t}} \le \overline{P^{\mathrm{PV}}_{\mathrm{m,r}}} \quad \forall \mathrm{r}, \forall \mathrm{m}, \forall t \tag{5.21}$$

$$P^{\mathrm{g}}_{\mathrm{m,i,t}} = P^{\mathrm{in}}_{\mathrm{m,i,t}}\Big|_{i=0} + \sum_{\mathrm{u}\in A^{u}_{i}} P^{\mathrm{DG}}_{\mathrm{m,g,t}} + \sum_{\mathrm{r}\in A^{r}_{i}} P^{\mathrm{PV}}_{\mathrm{m,r,t}} + \sum_{\mathrm{m'}\in A^{m}_{i}} P^{\mathrm{P2P}}_{\mathrm{m,m',t}}\Big|_{m\in A^{m'}_{m}} \quad \forall m, \forall i, \forall t \tag{5.22}$$

$$Q^{\mathrm{g}}_{\mathrm{m,i,t}} = Q^{\mathrm{in}}_{\mathrm{m,i,t}}\Big|_{i=0} + \sum_{\mathrm{u}\in A^{u}_{i}} Q^{\mathrm{DG}}_{\mathrm{m,g,t}} \quad \forall m, \forall i, \forall t \tag{5.23}$$

$$P^{\mathrm{d}}_{\mathrm{m,i,t}} = \sum_{d\in A^{d}_{i}} P^{\mathrm{Load}}_{\mathrm{m,d,t}} + \sum_{e\in A^{e}_{i}} P^{\mathrm{EB}}_{\mathrm{m,e,t}} \quad \forall m, \forall i, \forall t \tag{5.24}$$

$$Q^{\mathrm{d}}_{\mathrm{m,i,t}} = \sum_{d\in A^{d}_{i}} Q^{\mathrm{Load}}_{\mathrm{m,d,t}} \quad \forall m, \forall i, \forall t \tag{5.25}$$

$$\sum_{g} H^{\mathrm{DG}}_{\mathrm{m,g,t}} + \sum_{\mathrm{b}} H^{\mathrm{GB}}_{\mathrm{m,b,t}} + \sum_{\mathrm{e}} H^{\mathrm{EB}}_{\mathrm{m,e,t}} = H^{\mathrm{Load}}_{\mathrm{m,t}} \quad \forall \mathrm{m}, \forall t \tag{5.26}$$

5.3.3 Bilateral Trading Problem Formulation of MENMGs

Let $P^{P2P}_{m,m'}$ be the set of power exchanged between prosumer m and its neighbors m', where m' is the set of neighbors of prosumer m, with whom it can trade in the transactive energy market. When $P^{P2P}_{m,m'}$ is positive, it means the prosumer is buying electric power, while $P^{P2P}_{m,m'}$ is negative, it means the prosumer is selling power to the transactive energy market. Generally, a P2P trading problem can be modeled by (5.27)–(5.29), where Eq. (5.27) expresses the objective function of the problem. As it is clear, the objective is to minimize the operating costs of the MENMG in electricity markets. The objective function given in (5.27) contains four terms. The first and second terms, respectively, show the costs of purchasing electric power and gas fuel from upstream networks. The operating cost of DGs in the microgrid is also modeled by the third term. The P2P trading cost of MENMG in the TEM is expressed by the fourth term. Equation (5.28) shows the connection of MENMGs with each other. The power limit exchanged between MENMGs is also modeled using Eq. (5.29).

$$OF = \min \sum_{t} \sum_{m} \left(C^{\text{in}}_{\text{m,t}} P^{\text{in}}_{\text{m,t}} + C^{\text{Gas}}_{\text{m,t}} G^{\text{Gas}}_{\text{m,t}} + \sum_{g} C^{\text{DG}}_{\text{m,g,t}} P^{\text{DG}}_{\text{m,g,t}} + \sum_{m'} C^{\text{P2P}}_{\text{m,m}',\text{t}} P^{\text{P2P}}_{\text{m,m}',\text{t}} \right) \tag{5.27}$$

$$P^{\text{P2P}}_{\text{m,m}',\text{t}} + P^{\text{P2P}}_{\text{m}',\text{m,t}} = 0 \quad \forall \text{m} \in A^{\text{m}}_{\text{m}'}, \forall \text{m}', \forall t \tag{5.28}$$

$$-\overline{P^{\text{P2P}}_{\text{m}',\text{m}}} \le P^{\text{P2P}}_{\text{m,m}',\text{t}} \le \overline{P^{\text{P2P}}_{\text{m}',\text{m}}} \quad \forall \text{m}, \forall \text{m}', \forall t \tag{5.29}$$

s.t. (5.5)–(5.26).

It should be noted that the mathematical model presented in this study is strongly convex and can be easily solved using conventional solvers. However, the proposed problem is modeled as a centralized approach, and the privacy of peers trading with each other is not respected. To address this challenge, a central coordinator is defined where prosumers can share their offers and bids in the TEM while maintaining privacy. Therefore, a distributed optimization problem to preserve the privacy of each peer in the TEM is modeled as follows.

As it is clear from Eq. (5.28), the power exchanged between prosumers is modeled in a coupled approach. Therefore, Eq. (5.28) can be decomposed into two separate subproblems using an auxiliary variable $\rho^{\text{P2P}}_{\text{m,m}',\text{t}}$ and then solved using F-ADMM algorithm. Therefore, the coupled Eq. (5.28) can be formulated as follows [24].

$$\rho^{\text{P2P}}_{\text{m,m}',\text{t}} = P^{\text{P2P}}_{\text{m,m}',\text{t}} \quad \forall \text{m} \in A^{\text{m}}_{\text{m}'}, \quad \forall \text{m}', \forall t \tag{5.30}$$

$$\rho^{\text{P2P}}_{\text{m,m}',\text{t}} + \rho^{\text{P2P}}_{\text{m}',\text{m,t}} = 0 \quad \forall \text{m} \in A^{\text{m}}_{\text{m}'}, \quad \forall \text{m}', \forall t \tag{5.31}$$

$$\rho^{\text{P2P}}_{\text{m}',\text{m,t}} = P^{\text{P2P}}_{\text{m}',\text{m,t}} \quad \forall \text{m} \in A^{\text{m}}_{\text{m}'}, \quad \forall \text{m}', \forall t \tag{5.32}$$

According to the above modeling, the augmented Lagrangian of Eq. (5.27) is modeled as follows. The last term of the equation expressing the penalty function of mismatch of electric power exchanged between peers.

$$L\left(P^{\text{P2P}}_{\text{m,m}',\text{t}}, \rho^{\text{P2P}}_{\text{m,m}',\text{t}}, \pi_{\text{m,m}',\text{t}}\right) = \min \sum_{t} \sum_{m} \left(\begin{array}{l} C^{\text{in}}_{\text{m,t}} P^{\text{in}}_{\text{m,t}} + C^{\text{Gas}}_{\text{m,t}} G^{\text{Gas}}_{\text{m,t}} + \sum_{g} C^{\text{DG}}_{\text{m,g,t}} P^{\text{DG}}_{\text{m,g,t}} + \sum_{m'} C^{\text{P2P}}_{\text{m,m}',\text{t}} P^{\text{P2P}}_{\text{m,m}',\text{t}} \\ \sum_{m'} \pi_{\text{m,m}',\text{t}} \left(\rho^{\text{P2P}}_{\text{m,m}',\text{t}} - P^{\text{P2P}}_{\text{m,m}',\text{t}} \right) + \frac{\eta}{2} \rho^{\text{P2P}}_{\text{m,m}',\text{t}} - P^{\text{P2P}\ 2}_{\text{m,m}',\text{t2}} \end{array} \right) \tag{5.33}$$

s.t. (5.5)–(5.26), where η is a well-definite positive penalty parameter for constraint (5.27). Equation (5.33) is decomposable and can be divided into two subproblems. The first subproblem will be solved for each of the microgrids and according to the fixed data $\rho^{\text{P2P}}_{\text{m,m}',\text{t}}$ as follows:

$$OF = \min_{x} \sum_{t} \left(\begin{array}{l} C^{\mathrm{in}}_{\mathrm{m,t}} P^{\mathrm{in}}_{\mathrm{m,t}} + C^{\mathrm{Gas}}_{\mathrm{m,t}} G^{\mathrm{Gas}}_{\mathrm{m,t}} + \sum_{g} C^{\mathrm{DG}}_{\mathrm{m,g,t}} P^{\mathrm{DG}}_{\mathrm{m,g,t}} + \sum_{m'} C^{\mathrm{P2P}}_{\mathrm{m,m',t}} P^{\mathrm{P2P}}_{\mathrm{m,m',t}} \\ \sum_{m'} \pi^{\mathrm{k}}_{\mathrm{m,m',t}} \left(\rho^{\mathrm{P2P,k}}_{\mathrm{m,m',t}} - P^{\mathrm{P2P}}_{\mathrm{m,m',t}} \right) + \frac{\eta}{2} \rho^{\mathrm{P2P,k}}_{\mathrm{m,m',t}} - P^{\mathrm{P2P}\ 2}_{\mathrm{m,m',t2}} \end{array} \right) \tag{5.34}$$

s.t. (5.5)–(5.26).

In the second subproblem, a coordinator for P2P trading of MENMGs is defined, which is implemented by updating $P^{\mathrm{P2P}}_{\mathrm{m,m',t}}$ based on the decision $\rho^{\mathrm{P2P}}_{\mathrm{m,m',t}}$ in the first subproblem and considering constraint (5.28).

$$OF_{z} = \min_{z} \sum_{t} \sum_{m} \left(\sum_{m'} \pi^{\mathrm{k}}_{\mathrm{m,m',t}} \left(\rho^{\mathrm{P2P}}_{\mathrm{m,m',t}} - P^{\mathrm{P2P,k+1}}_{\mathrm{m,m',t}} \right) + \frac{\eta}{2} \rho^{\mathrm{P2P}}_{\mathrm{m,m',t}} - P^{\mathrm{P2P,k+1}2}_{\mathrm{m,m',t}\ \ 2} \right) \tag{5.35}$$

s.t. (5.28).

5.4 Fast ADMM Algorithm

After decomposing Eq. (5.33) into a main problem (5.34) and a subproblem (5.35), a decentralized optimization approach using a powerful distributed algorithm is needed to reach equilibrium. In this study, the powerful F-ADMM algorithm is used to achieve a strong equilibrium point. Generally, problem (5.34) is solved for each of the prosumer. Problem (5.35) also plays the role of a prosumer coordinator for P2P trading in transactive energy markets. According to Fig. 5.4, the details of solving the problem step by step are as follows:

1. Initialization of $\rho^{\mathrm{P2P}}_{\mathrm{m,m',t}}, \pi_{\mathrm{m,m',t}}$
2. Solve the main problem of prosumers using Eq. (5.34)
   (a) Update the variable $P^{\mathrm{P2P}}_{\mathrm{m,m',t}}$ and submit to the prosumer coordinator
3. Solve the subproblem (prosumer coordinator) using Eq. (5.35)
   (a) Update the variable $\rho^{\mathrm{P2P}}_{\mathrm{m,m',t}}$ and submit it to each of the prosumers
4. Check the convergence criterion $\delta = \rho^{\mathrm{P2P,k}}_{\mathrm{m,m',t}} - P^{\mathrm{P2P,k+1}}_{\mathrm{m,m',t}} \leq \varepsilon$
   (a) if the convergence criterion is satisfied, the problem is completed; otherwise, the problem is fed back to stage 1

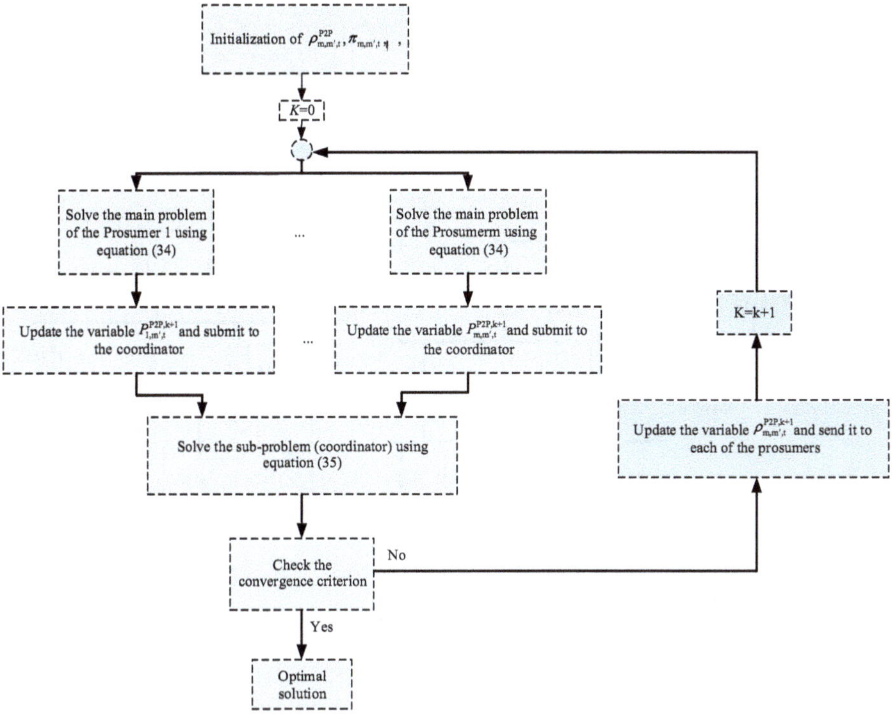

Fig. 5.4 The F-ADMM algorithm

5.5 Simulation and Numerical Results

In this study, a real system of 14-bus is used to model MENMG. Data and parameters related to 14-bus MENMG in [25] are available. It should be noted that the structure of all six MENMGs is the same as Fig. 5.5. As it is known, the multi-energy microgrid is equipped with a microturbine and a CHP, whose data is presented in Table 5.2. Additionally, microgrids 2, 4, and 6 are equipped with photovoltaic systems, whose forecasted production power is shown in Fig. 5.6. The peer-to-peer connections of the six prosumers are shown in Fig. 5.7.

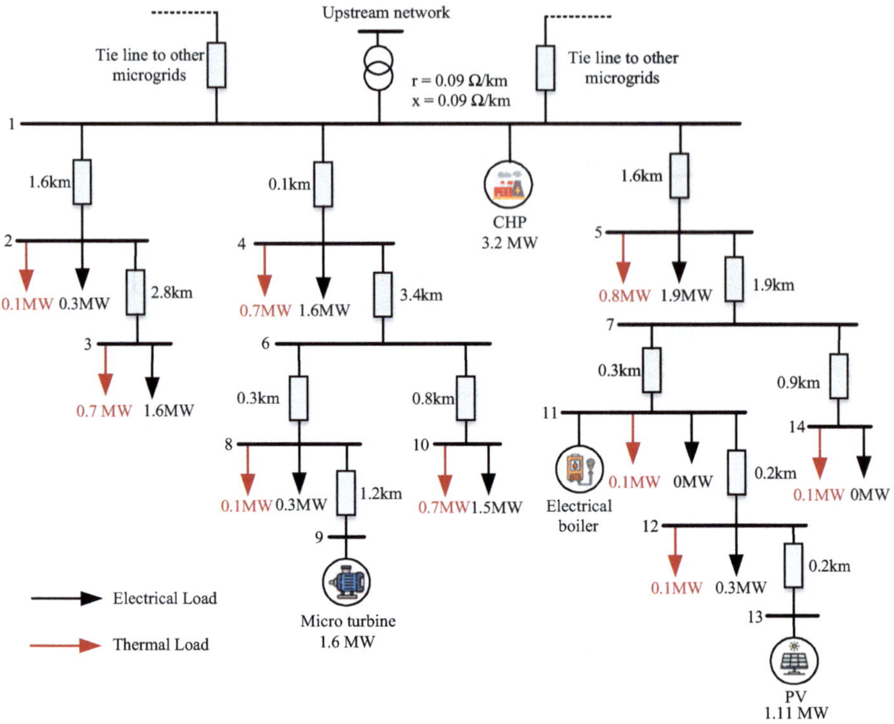

Fig. 5.5 14 nodes diagram of MENMG

Table 5.2 Data and parameters in the proposed problem [25]

Parameter	Unit	Value	Parameter	Unit	Value
$\underline{P}_{g}^{DG}, \overline{P_{g}^{DG}}$	MW	0, 1.2	$\underline{P}_{t}^{in}, \overline{P_{t}^{in}}$	MW	−20, 20
$\underline{Q}_{g}^{DG}, \overline{Q_{g}^{DG}}$	Mvar	0, 0.7	$\underline{Q}_{t}^{in}, \overline{Q_{t}^{in}}$	Mvar	−10, 10
$\eta_{m,b}^{GB}$	%	0.7	$-\overline{P_{m',m}^{P2P}}, \overline{P_{m',m}^{P2P}}$	MW	−3, 3
$\eta_{m,e}^{e}$	%	0.9	η	–	1000
$\gamma_{g,m}^{P}, \gamma_{g,m}^{H}$	–	2.4, 0.31	$C_{m,t}^{Gas}$	$/kcf	3
$R_{m,g}^{UP}, R_{m,g}^{DN}$	MW/h	1.5	$C_{m,g,t}^{DG\in MT}, C_{m,g,t}^{DG\in chp}$	$/MWh	65, 68
P_{m}^{R}	MW	3, 2.8, 2, 2.2	H_{m}^{R}	MW	0, 2.6, 1.8, 0

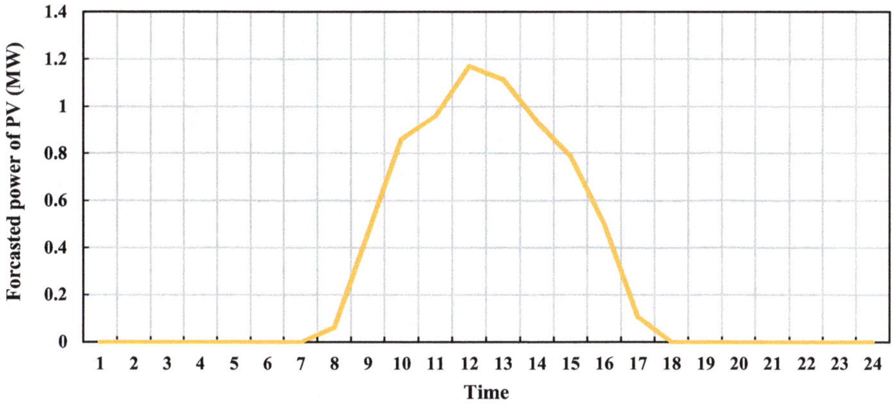

Fig. 5.6 Forecasted production power

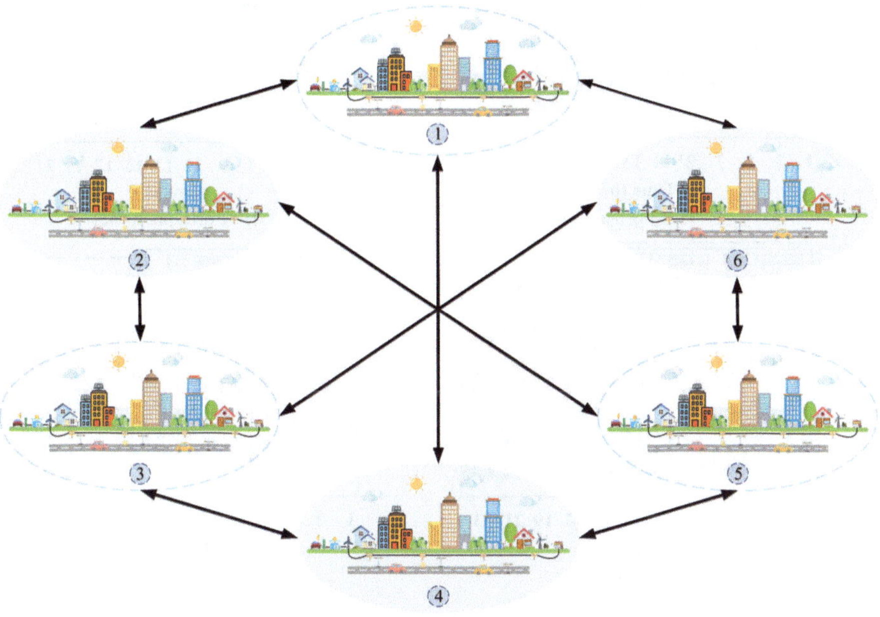

Fig. 5.7 The P2P connections of the six prosumers

5.6 Simulation and Results

The optimal scheduling of MENMGs load supply is illustrated in Fig. 5.8. As is well known, all MENMGs, the instantaneous equilibrium between supply and demand remains consistent throughout the studied hours. In the first prosumer, it supplies 50.08% and 28.34% of the local electricity market (LEM) and MT,

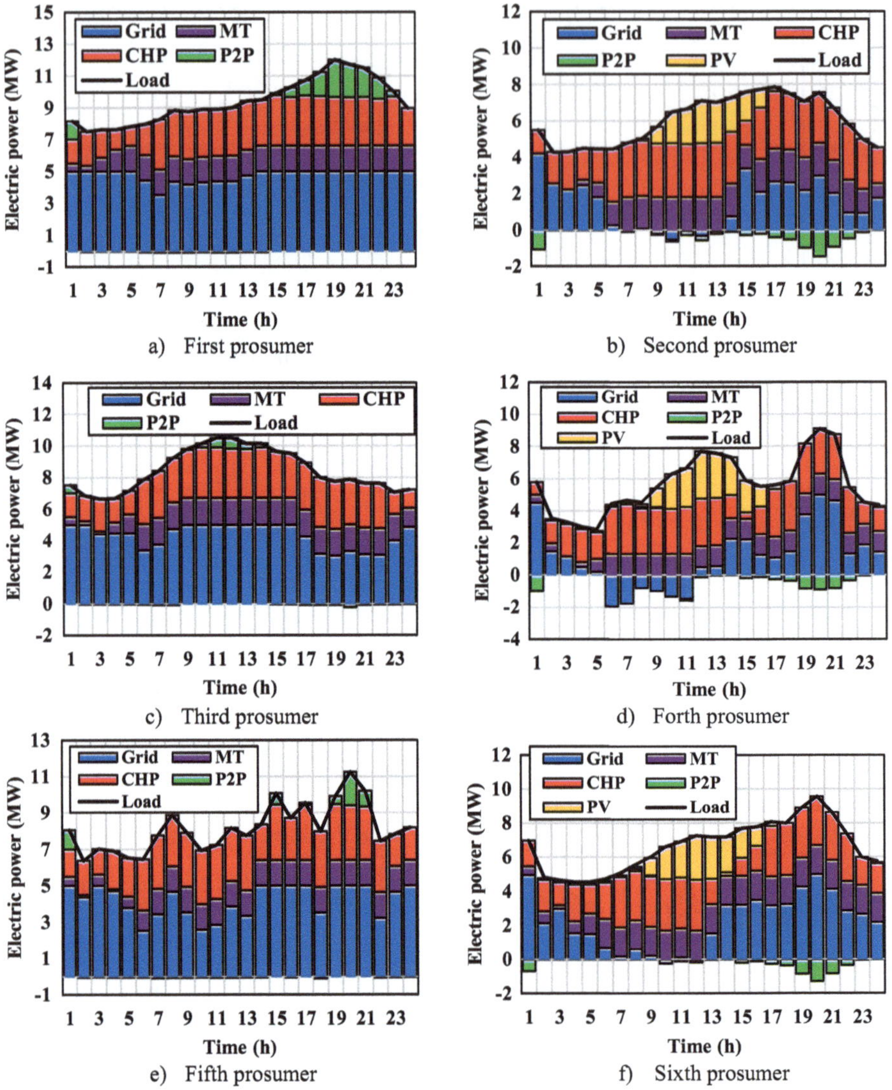

Fig. 5.8 Hourly scheduling of supplied electric demand of prosumers. (**a**) First prosumer, (**b**) second prosumer, (**c**) third prosumer, (**d**) forth prosumer, (**e**) Fifth prosumer, (**f**) sixth prosumer

respectively. On the other hand, 15.545% and 5.08% of electric loads are supplied by CHP and P2P trading. Interestingly, the first prosumer has appeared more as a power consumer (buyer) in the trading of TEMs. The second prosumer supplies 23.79% and 22.667% of its electricity loads by the LEM and MT, respectively. Moreover, CHP and PV units provide 42.99% and 9.63% of electrical loads, respectively. The second prosumer has emerged mostly as a producer (seller) in P2P energy trading. Additionally, the second prosumer shares its surplus production power in the LEMs during the hours of 9–13 (peak hours of PV production power). In the third prosumer, the breakdown of electric load supply is as follows: the LEM supplies 51.25%, MT contributes 17.07%, CHP provides 30.33%, and TEM fulfills 0.71% of the demand. In general, the third prosumer plays the role of a purchaser in energy markets. In the fourth prosumer, 19.52%, 19.69%, 40.66%, and 12.81% of electric loads are supplied by the LEM, MT, CHP, and PV, respectively. The fourth prosumer appears as the role of energy seller in electricity markets. In addition, the fourth prosumer significantly shares its excess production capacity in the LEM between 6 and 11 h. In the fifth prosumer, the electric load dispatched within the microgrid is as follows: the LEM provides 51.72%, MT contributes 14.48%, CHP supplies 31.28%, and TEM fulfills 1.19% of the demand. So the fifth prosumer acts more like a consumer. Finally, in the sixth prosumer, the LEM, MT, CHP, and PV supplies 33.6%, 22.3%, 33.4, and 10.6% of the microgrid electrical loads, respectively. As it is known, the sixth prosumers play the role of seller in electric power markets. Generally, paired prosumers (due to more generating power sources) can share more power and even maximize their profits by arbitrage operations between LEM and TEM. On the other hand, individual prosumers appear mostly in the role of buyers in electrical markets due to limited production power sources.

Figure 5.9 shows the optimal scheduling of MENMG heating load supply. As it is known, the heating loads are supplied in all MENMGs and the entire time period of the study. For example, in the first MENMG, 68.7% and 31.28% of heating loads are provided by CHP and gas boiler, respectively. In the second MENMG, CHP units and gas boilers are respectively responsible for supplying 76.1% and 23.9% of the heating loads of the multi-carrier microgrid. In the third MENMG, 68.14%, 30.93%, and 0.9% of heating loads are provided by CHP, gas boiler, and electric boiler, respectively.

Figure 5.10 shows the convergence proof of the total operating costs of prosumers. As can be seen, the prosumers achieve convergence after 20 iterations. In addition, Fig. 5.11 displays the convergence of the penalty term δ. It is observed that the δ-criterion of the 20th iteration achieves convergence. Figures 5.12, 5.13, and 5.14 show the P2P power trading of the first, fourth, and sixth prosumers with other peers, respectively. It's notable that power matches are evident in all P2P transactions between prosumers.

Table 5.3 shows the details of the operating costs of each prosumer in the energy market transactions. It's understood that the total operating costs of the first, second, and third prosumers amount to $17,640.1, $8867.367, and $15,570.21 per MW, respectively. Moreover, $7677.604, $15149.07, and $10427.63 per MWh are the operating costs of the fourth, fifth, and sixth prosumers. As it is known, the highest and lowest total operating costs are related to the first and fourth prosumers, respectively.

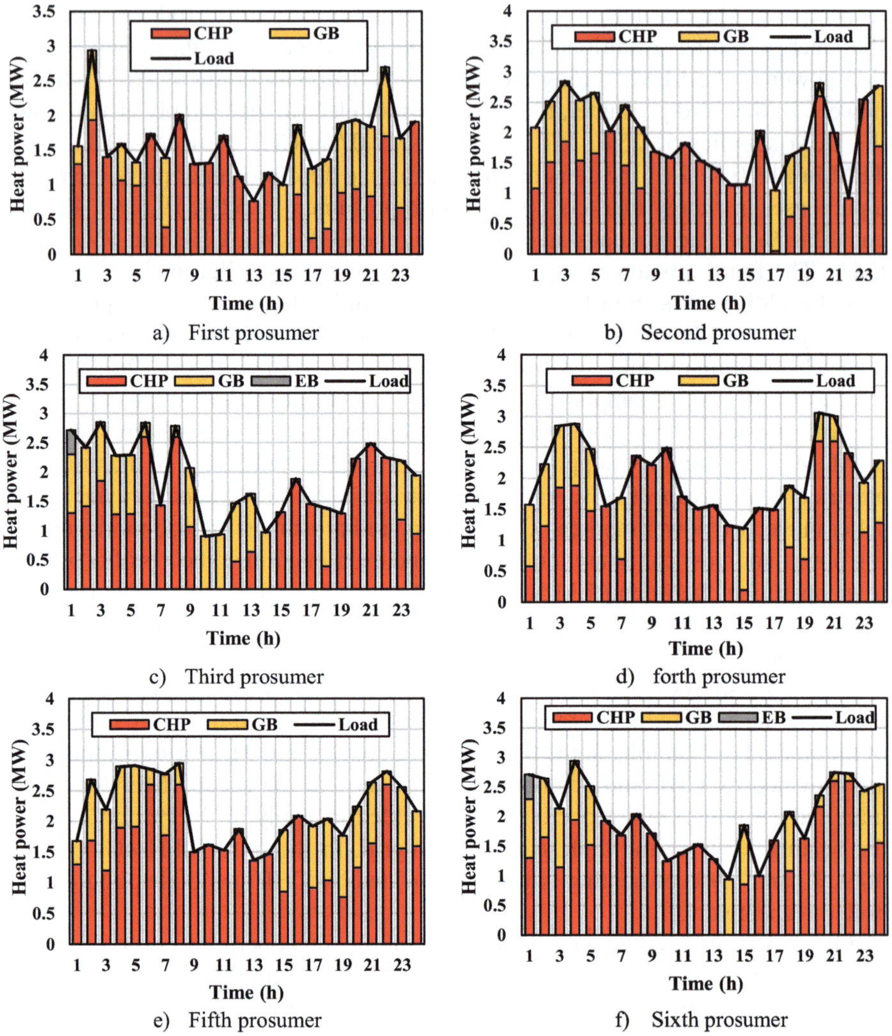

Fig. 5.9 Hourly scheduling of supplied heat demand of prosumers. (**a**) First prosumer, (**b**) second prosumer, (**c**) third prosumer, (**d**) forth prosumer, (**e**) Fifth prosumer, (**f**) sixth prosumer

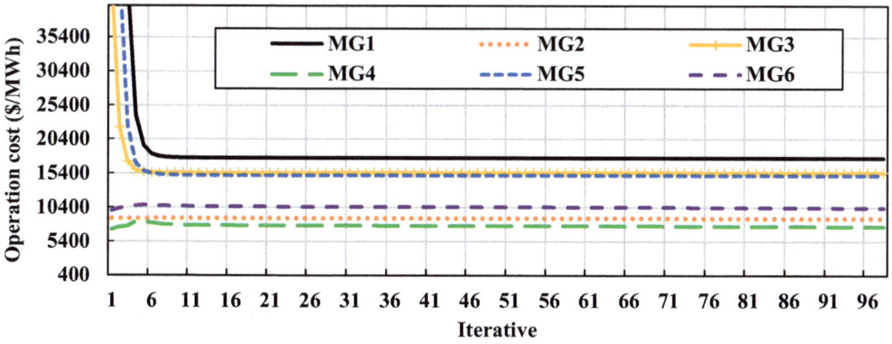

Fig. 5.10 Proof of convergence of operating costs of microgrids

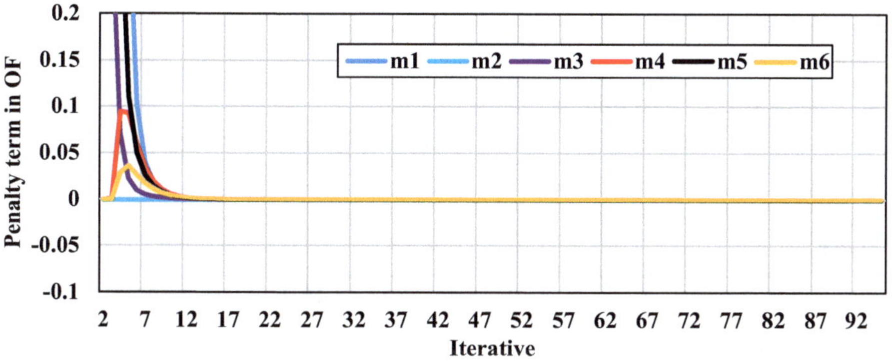

Fig. 5.11 Proof of convergence of ADMM algorithm penalty function

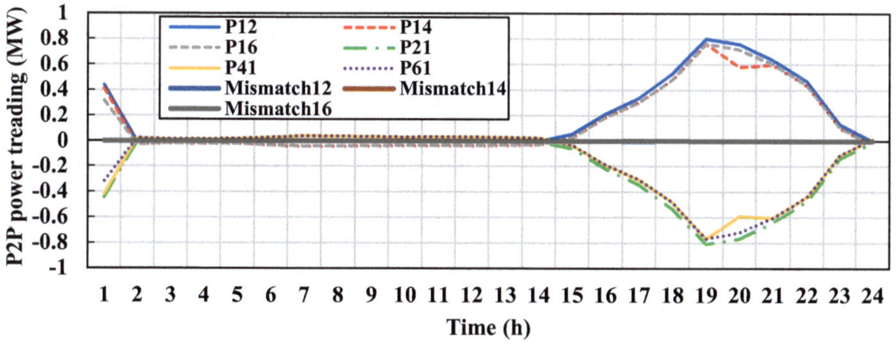

Fig. 5.12 First prosumer peer-to-peer trading with its neighbors

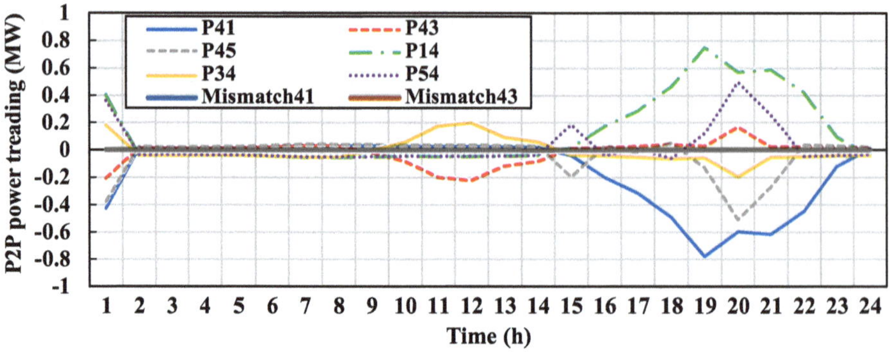

Fig. 5.13 Second prosumer peer-to-peer trading with its neighbors

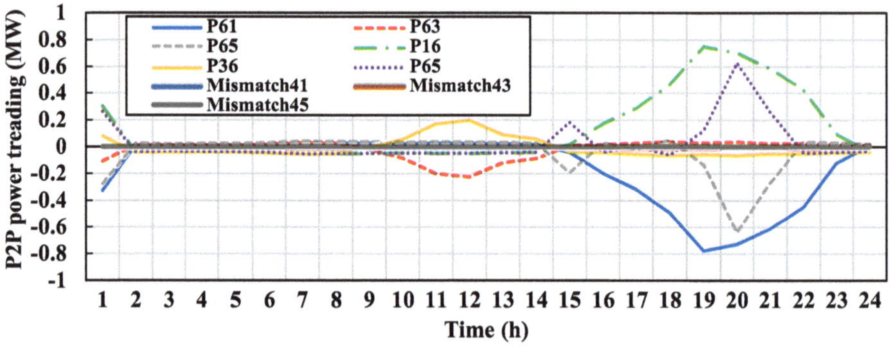

Fig. 5.14 Sixth prosumers peer-to-peer trading with its neighbors

Table 5.3 Operation costs of prosumers

	Total cost ($/MWh)	DG cost ($/MWh)	Gas cost ($/MWh)	Grid cost ($/MWh)	P2P cost ($/MWh)
MG1	17,640.41	6997.738	536.818	9182.296	922.6858
MG2	8867.363	6271.221	528.86	2656.128	−589.056
MG3	15,570.21	6644.788	533.0688	8259.693	132.1604
MG4	7677.604	5480.713	482.9796	2095.678	−384.804
MG5	15,159.07	6205.849	534.8934	8079.029	338.7782
MG6	10,427.63	6066.104	463.0109	4303.669	−408.235

5.7 Conclusion

In this study, a decentralized optimization approach was presented for trading multi-energy networked microgrids (MENMGs) in transactive energy markets (TEMs) and local electricity market (LEM). A real system test of 14 nodes was used to model the MENMG. The Distflow method was used to model the MENMG considering the physical constraints of the microgrid. In this study, it was assumed that microgrids are supplied through multi-energy distributed generation sources, LEMs and TEMs. To model energy trade on a peer-to-peer (P2P) basis, the fast alternating direction method of multipliers (F-ADMM) algorithm was employed. The obtained results show that the proposed structure can be effectively used for the transactions of multi-carrier microgrids, prosumers, and energy hubs in the framework of transactive energy concepts.

Appendix

The Disflow method was originally proposed to study distribution networks [26], which mainly work with radial topologies. The value of the bus voltage remains a variable in the branch power flow method, as it is a crucial indicator of power quality and is carefully monitored in distribution systems.

Let us consider a branch with impedance $z_{ij} = r_{ij} + jx_{ij}$ between buses i and j, where the complex voltages of the buses are obtained by $\dot{V}_i = V_i\angle\theta_i$ and $\dot{V}_j = V_j\angle\theta_j$, respectively. The complex power $\dot{S}_{ij}$ delivered in the main bus i can be expressed as:

$$\dot{S}_{ij} = P_{ij} + jQ_{ij} = \dot{V}_{ij}I_{ij} = \frac{\dot{V}_{ij}\left(\hat{V}_i - \hat{V}_j\right)}{r_{ij} - jx_{ij}} \tag{5.36}$$

where $\dot{V}_i\hat{V}_i = V_i^2$, $\dot{V}_i\hat{V}_j = V_iV_j\angle\theta_{ij}$, and $\theta_{ij} = \theta_i - \theta_j$, by multiplying the sides of the above equation by $r_{ij} - jx_{ij}$, equations become as follows:

$$P_{ij}r_{ij} + Q_{ij}x_{ij} - \left(Q_{ij}r_{ij} - P_{ij}x_{ij}\right) = V_i^2 - V_iV_j\left(\cos\theta_{ij} + j\sin\theta_{ij}\right) \tag{5.37}$$

By equalizing the imaginary part and the real part together, we have:

$$\begin{aligned} V_iV_j\cos\theta_{ij} &= V_i^2 - \left(P_{ij}r_{ij} + Q_{ij}x_{ij}\right) \\ V_iV_j\sin\theta_{ij} &= P_{ij}x_{ij} - Q_{ij}r_{ij} \end{aligned} \tag{5.38}$$

First, we raise both above equations to the power of two and then add them together. As a result of this operation, θ_{ij} is removed from the equation and the result of this operation will be as follows:

$$V_i^2V_j^2 = V_i^4 - 2V_i^2\left(P_{ij}r_{ij} + Q_{ij}x_{ij}\right) + \left(P_{ij}r_{ij} + Q_{ij}x_{ij}\right)^2 + \left(P_{ij}r_{ij} - Q_{ij}x_{ij}\right)^2 \tag{5.39}$$

where $z_{ij}^2 = r_{ij}^2 + x_{ij}^2$ is, then we have:

$$V_j^2 = V_i^2 - 2\left(P_{ij}r_{ij} + Q_{ij}x_{ij}\right) + z_{ij}^2\frac{\left(P_{ij}^2 + Q_{ij}^2\right)}{V_i^2} \tag{5.40}$$

$$I_{ij} = \frac{\left(P_{ij}\right)^2 + \left(Q_{ij}\right)^2}{V_i^2} \tag{5.41}$$

References

1. S. Chandak, P.K. Rout, The implementation framework of a microgrid: A review. Int. J. Energy Res. **45**(3), 3523–3547 (2021)
2. A. Muhtadi, D. Pandit, N. Nguyen, J. Mitra, Distributed energy resources based microgrid: Review of architecture, control, and reliability. IEEE Trans. Ind. Appl. **57**(3), 2223–2235 (2021)
3. M.A. Bidgoli, A. Ahmadian, Multi-stage optimal scheduling of multi-microgrids using deep-learning artificial neural network and cooperative game approach. Energy **239**, 122036 (2022)
4. A. Singh, B.S. Surjan, Microgrid: A review. Int. J. Res. Eng. Technol. **3**(2), 185–198 (2014)
5. G. Shahgholian, A brief review on microgrids: Operation, applications, modeling, and control. Int. Trans. Electr. Energy Syst. **31**(6), e12885 (2021)
6. S.A. Alavi, A. Ahmadian, M. Aliakbar-Golkar, Optimal probabilistic energy management in a typical micro-grid based-on robust optimization and point estimate method. Energy Convers. Manag. **95**, 314–325 (2015)
7. E.A. Soto, L.B. Bosman, E. Wollega, W.D. Leon-Salas, Peer-to-peer energy trading: A review of the literature. Appl. Energy **283**, 116268 (2021)
8. A. Aminlou, M.M. Hayati, K. Zare, Local Peer-to-Peer Energy Trading Evaluation in Micro-Grids with Centralized Approach, in *2023 8th International Conference on Technology and Energy Management (ICTEM)*, (Piscataway, IEEE, 2023), pp. 1–6
9. G.S. Thirunavukkarasu, M. Seyedmahmoudian, E. Jamei, B. Horan, S. Mekhilef, A. Stojcevski, Role of optimization techniques in microgrid energy management systems—A review. Energ. Strat. Rev. **43**, 100899 (2022)
10. M. Javidsharifi, H. Pourroshanfekr Arabani, T. Kerekes, D. Sera, J.M. Guerrero, Stochastic optimal strategy for power management in interconnected multi-microgrid systems. Electronics **11**(9), 1424 (2022)
11. C. Li, Z. Kang, H. Yu, H. Wang, K. Li, Research on energy optimization method of multi-microgrid system based on the cooperative game theory. J. Electr. Eng. Technol. **19**, 1–10 (2024)
12. A. Jani, S. Jadid, Two-stage energy scheduling framework for multi-microgrid system in market environment. Appl. Energy **336**, 120683 (2023)
13. H. Masrur, H. Khaloie, A.T. Al-Awami, S. El Ferik, T. Senjyu, Cost-aware modeling and operation of interconnected multi-energy microgrids considering environmental and resilience impact. Appl. Energy **356**, 122320 (2024)
14. X. Zhong, W. Zhong, Y. Liu, C. Yang, S. Xie, Optimal energy management for multi-energy multi-microgrid networks considering carbon emission limitations. Energy **246**, 123428 (2022)
15. T. Hai, J. Zhou, T. Muranaka, Management of renewable-based multi-energy microgrids with energy storage and integrated electric vehicles considering uncertainties. J. Energy Storage **60**, 106582 (2023)
16. H. Karimi, S. Jadid, S. Hasanzadeh, Optimal-sustainable multi-energy management of microgrid systems considering integration of renewable energy resources: A multi-layer four-objective optimization. Sustainable Product. Consumption **36**, 126–138 (2023)
17. C. Carvallo, F. Jalil-Vega, R. Moreno, A multi-energy multi-microgrid system planning model for decarbonisation and decontamination of isolated systems. Appl. Energy **343**, 121143 (2023)
18. P. Garavaso, F. Bignucolo, R. Turri, C. Zhang, Optimal Operation of Multi-Energy Microgrids Considering Network Constraints and Multiple Uncertainties, in *2021 IEEE 6th international forum on research and Technology for Society and Industry (RTSI)*, (IEEE, Piscataway, 2021), pp. 237–242
19. J. Gao, Z. Shao, F. Chen, Y. Chen, Y. Lin, H. Deng, Distributed robust operation strategy of multi-microgrid based on peer-to-peer multi-energy trading. IET Energy Syst. Integr. **5**(4), 376–392 (2023)
20. H. Liu, J. Li, S. Ge, X. He, F. Li, C. Gu, Distributed day-ahead peer-to-peer trading for multi-microgrid systems in active distribution networks. IEEE Access **8**, 66961–66976 (2020)
21. A.M. Saatloo, M.A. Mirzaei, B. Mohammadi-Ivatloo, A robust decentralized peer-to-peer energy trading in community of flexible microgrids. IEEE Syst. J. **17**(1), 640–651 (2022)

22. Z.-P. Yuan, P. Li, Z.-L. Li, J. Xia, Data-driven risk-adjusted robust energy management for microgrids integrating demand response aggregator and renewable energies. IEEE Trans. Smart Grid **14**(1), 365–377 (2022)
23. X. Chen, J. Zhai, Y. Jiang, C. Ni, S. Wang, P. Nimmegeers, Decentralized coordination between active distribution network and multi-microgrids through a fast decentralized adjustable robust operation framework. Sustainable Energy Grids Netw. **34**, 101068 (2023)
24. M.H. Ullah, J.-D. Park, Peer-to-peer energy trading in transactive markets considering physical network constraints. IEEE Trans. Smart Grid **12**(4), 3390–3403 (2021)
25. N. Nasiri, S. Zeynali, S.N. Ravadanegh, A tactical transactive energy scheduling for the electric vehicle-integrated networked microgrids. Sustain. Cities Soc. **83**, 103943 (2022)
26. W. Wei, J. Wang, *Modeling and Optimization of Interdependent Energy Infrastructures* (Springer, Cham, 2020)

Chapter 6
A Cost-Emission Trade-offs in Hybrid Energy Systems Using a Multi-Objective Approach with Demand Response Integration

Mehran Ghahramani, Mehrdad Ghahramani, and Mehdi Abapour

6.1 Introduction

Today, the transition to renewable energies and the reduction in the use of fossil fuels are essential for advancing a sustainable city by decreasing environmental pollution. Renewable energy sources generate electricity without emitting carbon footprint pollutants, which are commonly released by burning fossil fuels, thereby supporting the goals of a sustainable city [1]. Continuing to use fossil fuels in the same manner will exacerbate global warming by releasing significant amounts of carbon-related materials and other greenhouse gases into the atmosphere [2]. Utilizing renewable energies in a distributed form within a sustainable city—such as solar energy, fuel cells, and energy storage—enables more localized energy generation, minimizes losses, and enhances efficiency. By integrating these systems into traditional power infrastructure and sustainable buildings, we can substantially reduce energy consumption and contribute to the development of a sustainable city, where resources are used more efficiently and environmental impacts are minimized [3].

In recent times, distributed generating technologies such as fuel cells [4], solar systems [5], and battery storage [6] have become increasingly popular for meeting energy demand alongside the traditional power supply provided by the upstream grid. Using these resources separately may not result in their optimal efficiency. Thus, the most efficient utilization of these resources may be achieved by operating them as a unified system, such as a hybrid energy system (HES) [7]. Furthermore,

M. Ghahramani (✉) · M. Abapour
Faculty of Electrical and Computer Engineering, University of Tabriz, Tabriz, Iran
e-mail: abapour@tabrizu.ac.ir

M. Ghahramani
School of Engineering, Edith Cowan University, Joondalup, WA, Australia
e-mail: m.ghahramani@ecu.edu.au

A. Ahmadian et al. (eds.), *Sustainable Energy Resources in Smart Cities*, Green Energy and Technology, https://doi.org/10.1007/978-3-031-91516-1_6

as the population continues to grow, there is an increasing need for energy, necessitating the construction of additional power plants to fulfill this demand. However, this expansion in power plant infrastructure also results in a corresponding increase in pollution levels. The need to address environmental pollution has made it imperative to limit the release of greenhouse gases, such as CO_2 [8]. The studies and research on HES are categorized into two categories including optimal sizing and operation. The optimal sizing of HESs has been a widely discussed subject among researchers. The study of [9] enhances the sizing of HESs through a Hybrid Grey Wolf with Cuckoo Search Optimization algorithm, yielding better cost and performance outcomes than other algorithms. Optimal sizing of hybrid energy systems (HES) ensures reliability, sustainability, and cost-efficiency. This process involves evaluating various components, methods, constraints, and optimization tools to achieve efficient energy production [10]. An evolutionary algorithm designed to optimize the sizing of hybrid energy storage systems was introduced in [11]. This approach enhances efficiency, extends battery life, and significantly improves energy loss and performance. The study of [12] introduces a multi-objective optimization model for sizing an HES that includes solar, wind, diesel, and battery components. The model aims to minimize both costs and carbon emissions effectively. The authors of [13] investigate the optimal sizing of hybrid solar, wind, and hydroelectric pumped storage systems in Egypt, employing the Whale Optimization Algorithm and other techniques to minimize energy costs effectively. Optimal sizing of HESs that incorporate power sharing and electric vehicles, utilizing Multi-Objective Particle Swarm Optimization and Multi-objective Crow Search algorithms to enhance the cost-effectiveness and reliability of the system, is suggested in [14]. In [15], hybrid renewable energy systems for Al-Karak governorate, Jordan, are evaluated, considering both off-grid and grid-connected options. It explores configurations to maximize renewable energy integration, minimize costs, and reduce CO_2 emissions. The optimal system integrates PV and wind power with grid connection and battery storage, ensuring reliable energy access, cost savings, and environmental benefits. A model for designing an independent HES with PV and wind sources, storage, and a diesel generator to optimize electricity costs and enhance reliability in rural areas was introduced in [16]. The study of [17] examines HESs in commercial buildings, focusing on PV, cooling storage, and batteries in China. It develops an optimization model to determine the best technology mix and financial benefits, highlighting cooling storage's economic advantages over batteries and the role of PV in enhancing system efficiency. The adoption of renewable energy sources and the need to optimize the sizing of HESs to enhance reliability and cost-efficiency are examined in [18]. This paper reviews recent classification, evaluation indicators, and sizing methodologies, emphasizing advancements in traditional and AI-based approaches for improved performance of HESs. Many researchers have analyzed and investigated the optimal operation of HESs. The work of [9] investigates optimizing a PV, wind, battery, and gas boiler; HES aims to meet electric and thermal demand for a remote community. The proposed system is analyzed for cost-effectiveness and environmental benefits using genetic algorithm techniques. The study of [10] introduces a two-stage robust scheduling model for the optimal operation of an HES, combining solar, wind, hydropower, thermal, and pumped storage sources in a competitive energy and reserve market. The model, tested on modified IEEE 5-bus and 24-bus systems, optimizes day-ahead and worst-case energy dispatch costs, incorporates uncertainty in wind speed, and utilizes demand response programs (DRPs) for load flexibility. A cost-efficient operation of an HES is evaluated in [11], which optimizes energy by shifting loads from peak to off-peak periods, minimizing total costs, and maximizing revenue from power exports with mixed-integer linear programming. The study of [12] evaluates a stand-alone HES for Uttara University in Bangladesh, focusing on cost-efficiency, technical performance, and environmental benefits. The optimal configuration of PV-WT-DG-BAT minimizes costs and emissions while maximizing renewable energy use and job creation. Sensitivity analysis showed that fuel price changes significantly impact costs, and the system's performance improves with increasing solar radiation. The study suggests adopting smart energy management and incentivizing local renewable manufacturing can enhance HES benefits. A novel hybrid DRP that combines time-of-use and incentive-based approaches, along with an HES consisting of PV panels, wind turbines, and energy storage, is introduced for Kanyakumari, India, in [13]. In [14], a novel day-ahead energy management strategy integrating DRP into PV, diesel, and battery HES for remote off-grid power systems in Northern Canada is introduced. By combining a probabilistic fuzzy DRP model with operational optimization, the framework aims to maximize benefits for utilities and consumers while minimizing human interaction and communication complexities. In [15], advancements in sizing and energy management of HES are reviewed, focusing on components like solar PV, WTs, batteries, and hydrogen storage. Optimization methods and tools for improving system performance are evaluated based on technical, reliability, environmental, and social factors in various locations and applications. The techno-economic feasibility of two HESs for electricity and hydrogen production in Maroua, Cameroon, is examined in [16]. The first scenario combines solar PV, fuel cells, and biogas, while the second adds battery storage. This analysis shows that integrating these components can offer a reliable alternative energy source, reduce reliance on traditional sources, and improve regional electricity access, with economic and environmental benefits. In [17], the research investigates the optimal integration and operation of a wind turbine, PV system with mechanical gravity energy storage, and electrochemical batteries. The study uses forecast models and an adaptive energy management strategy to improve design accuracy and operational efficiency. In reference [18], the study identifies the lowest energy cost for renewable systems at five global locations by analyzing power generation and hydrogen production using solar and wind resources. The research finds that configurations including PV and WT, along with BAT and hydrogen tanks, are cost-effective. This chapter examines the cost-emission operation of a PV-FC-BAT HES. The study considers the impact of DRP. A multi-objective problem with two contradictory objectives has been suggested in this chapter. The suggested model is resolved with the weighted sum methodology. The methodology used yields several solutions for the given model; however, the desired solution is the one that can achieve a balance between the opposing goal functions. To accomplish this, a fuzzy constraint satisfaction strategy is employed to choose the optimal solution among the solutions that have been obtained. Additionally, the implementation of DRP aims to flatten the load curve by redistributing a portion of the demand from peak periods to other periods. This strategy helps to minimize the overall cost of the system.

The remaining sections of the chapter are classified in the following order: Sect. 6.2 outlines the methodologies and strategies for resolving multi-objective models. Section 6.3 presents the formulation of a cost and emission optimization model for an on-grid PV-FC-BAT HES. Section 6.4 examines two study cases and compares the resulting outcomes. Section 6.5 presents the results and conclusions.

6.2 Multi-objective Modeling

Various methodologies and methods have been developed to address the challenge of solving multi-objective optimization. The strategies that may be used to address the multi-objective model include the e-constraint method [19], evolutionary algorithms [20], and the weighted sum approach [21]. Each solution has distinct characteristics and employs specific processes to address multi-objective issues. The multi-objective optimization model suggested in this study is solved using the weighted sum technique. This strategy involves obtaining many Pareto optimal solutions by applying weights to competing goal functions and selecting the result from the optimal Pareto group, considering the varying weights and values. The objective function of a multi-objective issue in the weighted sum approach is expressed as follows:

$$\min\left[\Phi\right] = w_1\Phi_1 + w_2\Phi_2 \tag{6.1}$$

$$w_1 + w_2 = 1 \tag{6.2}$$

Since the objectives employed in models often have various dimensions, they need to be translated to values per unit. This is accomplished by using the fuzzy satisfactory technique, which is covered in more detail in the following section.

6.2.1 *Fuzzy Satisfying Method*

Various techniques may be used to address the multi-objective issue, resulting in a multitude of optimum solutions. The optimal solution we anticipate can achieve a mutually beneficial strategy between the objectives. The first step in a multi-objective optimization model is to normalize the objective functions since they may have varying dimensions and ranges. Let's examine a case where there are N objectives that require minimization. Each objective function is assigned a fuzzy membership that translates it to the range zero to one. The linear function of membership for the nth problem of the ith objective is calculated as follows [22]:

$$\Phi_i^n = \begin{cases} 1 & \Phi_i^n \leqslant \Phi_i^{\min} \\ \dfrac{\Phi_i^n - \Phi_i^{\max}}{\Phi_i^{\min} - \Phi_i^{\max}} & \Phi_i^{\min} \leqslant \Phi_i^n \leqslant \Phi_i^{\max} \\ 0 & \Phi_i^n \geqslant \Phi_i^{\max} \end{cases} \tag{6.3}$$

The highest and lowest levels of the objective i in the generated outcomes of the multi-objective issue are denoted by $\Phi_i^{\max}$ and $\Phi_i^{\min}$ in Eq. (6.3). The optimum result of the nth problem is expressed as Φ_i^n. The objective functions are normalized as stated in Eqs. (6.4) and (6.5), respectively.

$$\mathrm{Cost}_{\mathrm{pu}} = \Phi_{1,\mathrm{pu}} = \frac{\mathrm{Cost} - \mathrm{Cost}^{\max}}{\mathrm{Cost}^{\min} - \mathrm{Cost}^{\max}} \tag{6.4}$$

$$\mathrm{Co}_{\mathrm{pu}} = \Phi_{2,\mathrm{pu}} = \frac{\mathrm{Co} - \mathrm{C}^{\max}}{\mathrm{Co}^{\min} - \mathrm{Co}^{\max}} \tag{6.5}$$

The membership function of the nth solution is computed using Eq. (6.6). The maximum weakest membership function (6.7) represents the optimal solution for the multi-objective optimization model.

$$\Phi^n = \min\left(\Phi_1^n, \ldots, \Phi_N^n\right) \qquad n = 1, \ldots, N_P \tag{6.6}$$

$$\mathrm{Co}_{\mathrm{pu}} = \Phi_{2,\mathrm{pu}} = \frac{\mathrm{Co} - \mathrm{C}^{\max}}{\mathrm{Co}^{\min} - \mathrm{Co}^{\max}} \tag{6.7}$$

The following is a concise presentation of the solution method's stages, along with the explanations provided above:

1. Minimize x while considering all equal and unequal restrictions. Determine the value of y, which is its maximum.
2. Minimize y while considering all equal and unequal limitations. Determine the value of x, which is the maximum.
3. The lowest and highest x and y amounts are computed in the first and second phases.
4. The following model is currently optimizing the suggested multi-objective formula employing the weighted sum technique:

$$\min\left[\Phi\right] = \mathrm{w}_1\Phi_1 + \mathrm{w}_2\Phi_2$$

s.t.

$$\begin{cases} w_1 + w_2 = 1 \\ \text{all of constraits} \end{cases}$$

5. Different weighted coefficients (w_1, w_2), where $w_1 + w_2 = 1$, lead to different answers. As a result, several Pareto answers are offered.
6. The fuzzy satisfactory approach chooses the optimal course of action. As a result, Eq. (6.31) converts the goals to per unit in various answers. The degree of optimality of the nth answer to the kth function is shown in $\propto_N^n$.
7. Here is how to determine the membership function of the nth solution:

$$\mu^n = \min\left(\mu_1^n, \ldots, \mu_N^n\right)$$
$$n = 1, \ldots, N_P$$

8. The solution with the highest weakest membership function is the optimal one.

$$\mu^{\max} = \max\left(\mu^1, \ldots, \mu^{Np}\right)$$

6.2.2 Problem Formulation

6.2.2.1 Cost Function

The primary function is to lower the system's total cost (6.8). The overall cost is calculated by adding the expenditures of purchasing gas and energy and subtracting the benefit of selling energy. The benefits and expenses described above are given separately in Eqs. (6.9)–(6.11).

$$\min \Phi 1 = \text{Cost}_{\text{purchase}}^{\text{grid}} + \text{Cost}^{\text{gas}} - \text{Revenue}^{\text{PV}} = \text{Cost}^{\text{Total}} \tag{6.8}$$

Equation (6.8) illustrates the total cost of HES which $\text{Cost}_{\text{purchase}}^{\text{grid}}$, Cost^{gas}, and $\text{Revenue}^{\text{PV}}$ represent the cost of purchased energy from upstream grid, purchased gas, and the revenue from selling power to the upstream grid, respectively.

$$\text{Cost}_{\text{purchase}}^{\text{grid}} = \sum_m P_{\text{base}}^{\text{grid}} + \sum_m \sum_h \left(E_{m,h,\text{charge}}^{\text{grid}} + E_{m,h,\text{load}}^{\text{grid}}\right) \cdot D \cdot P_{m,h}^{\text{grid}} \tag{6.9}$$

Equation (6.9) states that the cost of obtaining electricity from the upstream grid equals the energy bought for a year multiplied by the energy's purchase price. Apart from the expenses incurred for each unit of energy consumed, a monthly base price is also required. $P_{\text{base}}^{\text{grid}}$, $E_{m,h,\text{charge}}^{\text{grid}}$, $E_{m,h,\text{load}}^{\text{grid}}$, and $P_{m,h}^{\text{grid}}$ show the recurring fixed payment for electricity purchased from the upstream grid, purchased electric power to charge battery, electricity acquired to meet demand, and the cost at which energy is acquired from the upstream grid, respectively. D also shows the days for each month.

$$\mathrm{Cost}^{\mathrm{gas}} = \sum_{m} P_{\mathrm{base}}^{\mathrm{grid}} + \sum_{m}\sum_{h}\left(\frac{E_{m,h,\mathrm{load}}^{fc} + E_{m,h,\mathrm{charge}}^{fc}}{\eta_{e}^{fc}} + \frac{H_{m,h}^{bb}}{\eta^{bb}}\right) \cdot D \cdot P_{m,h}^{\mathrm{gas}} \tag{6.10}$$

Equation (6.10), which shows how much gas is used by the FC and backup burner multiplied by the gas price, represents the cost of gas bought. It should be mentioned that a base fee has to be paid each month in addition to the cost for each quantity of gas used. $E_{m,h,\mathrm{load}}^{fc}$ and $E_{m,h,\mathrm{charge}}^{fc}$ represent the electricity produced by the FC to meet demand and to charge battery. $H_{m,h}^{bb}$ shows the heat generated by backup burner. η_{e}^{fc} and η^{bb} also represent the electrical efficiency of FC and the efficiency of backup burner. $P_{m,h}^{gas}$ shows the cost at which city gas is obtained in the HES.

$$\mathrm{Revenue}^{\mathrm{PV}} = \sum_{m}\sum_{h} E_{m,h,\mathrm{sale}}^{\mathrm{PV}} \bullet D \bullet P_{\mathrm{sale}}^{\mathrm{PV}} \tag{6.11}$$

The final part of the initial objective function, which indicates the profit gained by selling surplus electricity generated by the PV system to the grid, is expressed by Eq. (6.11). This equation states that the extra electricity from the PV is transferred to the upstream grid at a predetermined price after a year, resulting in a total cost decrease. $E_{m,h,\mathrm{sale}}^{PV}$ represents the sold power by PV system, and P_{sale}^{PV} also indicates the cost at which surplus PV electricity is transferred to the upstream grid.

6.2.2.2 Emission Function

In this study, we have attempted to address environmental concerns such as CO_2 emissions while minimizing the overall cost of the system. Reducing CO_2 emissions is thus the second function of the suggested article (6.12). Equations (6.13) and (6.14) express each of the terms of the second objective function separately.

$$\min \Phi 2 = \mathrm{CO}^{\mathrm{total}} = \mathrm{CO}^{\mathrm{gas}} + \mathrm{CO}^{\mathrm{grid}} \tag{6.12}$$

$$\mathrm{CO}^{\mathrm{gas}} = \sum_{m}\sum_{h}\left(\frac{E_{m,h,\mathrm{load}}^{fc} + E_{m,h,\mathrm{charge}}^{fc}}{\eta_{e}^{fc}} + \frac{H_{m,h}^{bb}}{\eta^{bb}}\right) \bullet D \cdot \mathrm{CI}^{\mathrm{gas}} \tag{6.13}$$

$$\mathrm{CO}^{\mathrm{grid}} = \sum_{m}\sum_{h}\left(E_{m,h,\mathrm{charge}}^{\mathrm{grid}} + E_{m,h,\mathrm{load}}^{\mathrm{grid}}\right) \bullet D \bullet \mathrm{CI}^{\mathrm{grid}} \tag{6.14}$$

Equation (6.13) states that the quantity of CO_2 released into the atmosphere as a result of gas usage is directly proportional to the carbon footprint of city gas multiplied by the amount of gas consumed throughout a year by FC and backup burner. Fossil fuels are used in the upstream grid's generation of electricity. Accordingly, the quantity of CO_2 released due to the electricity supplied by the upstream grid is equal to the grid's carbon intensity times the power bought from it for a year.

6.2.2.3 Thermal Load

This section expresses the pertinent limitations and the method by which the thermal load is delivered Eqs. (6.15)–(6.23). The thermal requirement in the suggested article is met by heat released from the thermal energy storage tank in Eq. (6.15). $H_{m,h}^{\text{demand}}$ and $H_{m,h,out}^{st}$ represent the thermal demand and the thermal energy discharged from the heat storage tank to meet the thermal load.

$$H_{m,h}^{\text{demand}} = H_{m,h,out}^{st} \tag{6.15}$$

FCs and backup burners provide heat, which charges the tank (6.16). After receiving the heat produced by the backup burner and FC, the storage tank releases the stored thermal energy at various intervals to meet the thermal load. $H_{m,h,in}^{st}$ shows the thermal energy generated by FC and backup burner to save in the storage tank. Also, $H_{m,h}^{fc}$ represents the thermal energy produced by FC.

$$H_{m,h,in}^{st} = H_{m,h}^{bb} + H_{m,h}^{fc} \tag{6.16}$$

The thermal energy generated by the FC and the backup burner is constrained by Eqs. (6.17) and (6.18), in that order. H_{capacity}^{bb} is the rated capacity of backup burner. HE_{ratio}^{fc} in Eq. (6.18) represents the thermal energy of FC to electricity ratio.

$$H_{m,h}^{bb} \leqslant H_{\text{capacity}}^{bb} \tag{6.17}$$

$$H_{m,h}^{fc} \leqslant \left(E_{m,h,\text{load}}^{fc} + E_{m,h,\text{charge}}^{fc}\right) \bullet HE_{\text{ratio}}^{fc} \tag{6.18}$$

The amount of heat available within the storage tank at any given time is determined by adding the thermal energy from the previous time to the current thermal energy, which subtracts the heat needed to provide thermal load. Equation (6.19) establishes the available heat, while Eq. (6.20) limits it.

$$H_{m,h}^{\text{st}} = (1-\zeta) \bullet H_{m,h-1}^{\text{st}} + H_{m,h,\text{in}}^{\text{st}} \bullet \eta_{\text{in}}^{\text{st}} - \left(H_{m,h,\text{out}}^{\text{st}} / \eta_{\text{out}}^{\text{st}}\right) \tag{6.19}$$

$$H_{m,h}^{\text{st}} \leqslant H_{\text{capacity}}^{\text{st}} \tag{6.20}$$

$H_{m,h}^{\text{st}}$ is the thermal energy present within the storage tank. ζ also reflects the thermal energy lost from the tank. $\eta_{\text{in}}^{\text{st}}$ and $\eta_{\text{out}}^{\text{st}}$ represent the charging and discharging efficiency of the storage tank, respectively. Equations (6.21) and (6.22) constrain the storage tank's charging and discharging rate. $H_{\text{capacity}}^{\text{st}}$ shows the nominal capacity of utilized the storage tank.

$$H_{m,h,\text{in}}^{\text{st}} \leqslant H_{\text{capacity}}^{\text{st}} \tag{6.21}$$

$$H_{m,h,\text{out}}^{\text{st}} \leqslant H_{\text{capacity}}^{\text{st}} \tag{6.22}$$

6.2.2.4 Electric Load

The suggested study incorporates a grid and an HES, which supplies the electrical load while considering the DRP equation (6.23).

$$E_{m,h,\mathrm{DRP}}^{\mathrm{demand}} = E_{m,h,\mathrm{load}}^{\mathrm{grid}} + E_{m,h,\mathrm{load}}^{\mathrm{pv}} + E_{m,h,\mathrm{load}}^{\mathrm{fc}} + E_{m,h,\mathrm{discharge}}^{\mathrm{bat}} \tag{6.23}$$

The energy acquired from the upstream grid to provide load is the first term in Eq. (6.23). $E_{m,h,\mathrm{DRP}}^{\mathrm{demand}}$, $E_{m,h,\mathrm{load}}^{\mathrm{grid}}$, $E_{m,h,\mathrm{load}}^{\mathrm{pv}}$, and $E_{m,h,\mathrm{load}}^{\mathrm{fc}}$ show the electrical load with considering DRP, the purchased power from upstream grid, the generated power by PV system, and the generated power by FC, which all of them for supplying the load, and $E_{m,h,\mathrm{discharge}}^{\mathrm{bat}}$ also represents the discharged power by battery to supply load. A portion of the energy bought is used to charge storage and powering loads. The overall energy required from the grid is constrained by Equation (6.24), which represents the aggregate of the energy needed for battery changing and fulfilling the load requirements. M is a constant number, and $u_{m,h,\mathrm{pur}}^{\mathrm{grid}}$ is a binary variable that it takes the value of 1 if energy is sourced from the upstream grid and 0 if it is not.

$$E_{m,h,\mathrm{charge}}^{\mathrm{grid}} + E_{m,h,\mathrm{load}}^{\mathrm{grid}} \leqslant M \bullet u_{m,h,\mathrm{pur}}^{\mathrm{grid}} \tag{6.24}$$

The electricity that PV systems produce is used for various purposes. Equation (6.25) states that generated energy may be sold to the grid during periods of surplus production and utilized to provide load and charge batteries. $E_{m,h,\mathrm{load}}^{\mathrm{pv}}$ and $E_{m,h,\mathrm{charge}}^{\mathrm{pv}}$ represent the power produced by PV to meet demand and charge battery, respectively. α and β denote the effectiveness of the PV and inverter, respectively. A represents the area allocated for installing the PV system, and $R_{m,\,h}$ represents solar radiation. There is a restriction of Eq. (6.26) on the energy sold to the grid. $u_{m,h,\mathrm{sale}}^{\mathrm{PV}}$ X is a binary variable that takes the value of 1 if surplus produced power by PV is transferred to the upstream grid, and 0 otherwise. It is important to remember that PV and the grid cannot interchange energy simultaneously, as represented by Eq. (6.27).

$$E_{m,h,\mathrm{load}}^{\mathrm{pv}} + E_{m,h,\mathrm{charge}}^{\mathrm{pv}} + E_{m,h,\mathrm{sale}}^{\mathrm{pv}} = \alpha \bullet \beta \bullet A \bullet R_{m,h} \tag{6.25}$$

$$E_{m,h,\mathrm{sale}}^{\mathrm{pv}} \leqslant M \bullet u_{m,h,\mathrm{sale}}^{\mathrm{PV}} \tag{6.26}$$

$$u_{m,h,\mathrm{pur}}^{\mathrm{grid}} + u_{m,h,\mathrm{sale}}^{\mathrm{PV}} \leqslant 1 \tag{6.27}$$

Electric power produced by FCs is the third term in Eq. (6.23). The FCs generate electricity to power the load and recharge the battery storage using the bought gas. The energy produced must not exceed the FC's rated capacity in Eq. (6.28). $E_{\mathrm{capacity}}^{\mathrm{fc}}$ is the rated capacity of FC.

$$E_{m,h,\mathrm{load}}^{\mathrm{fc}} + E_{m,h,\mathrm{charge}}^{\mathrm{fc}} \leqslant E_{\mathrm{capacity}}^{\mathrm{fc}} \tag{6.28}$$

The power discharged by battery storage is the fourth term in Eq. (6.16). FCs, PV systems, and the upstream grid all charge batteries in Eq. (6.29). It should be remembered that the energy entering the storage must not be more than what the battery can hold during charging. N is a constant number, and $u_{m,h,\text{charge}}^{\text{bat}}$ represents a binary variable that if the battery is in charging mode, its value is equal to 1; otherwise, it is equal to zero.

$$\left(E_{m,h,\text{charge}}^{\text{grid}} + E_{m,h,\text{charge}}^{\text{fc}} + E_{m,h,\text{charge}}^{\text{pv}}\right) \leqslant N \cdot u_{m,h,\text{charge}}^{\text{bat}} \tag{6.29}$$

The energy that batteries store is preserved from several sources. Every hour, the energy that is accessible inside the battery storage is determined by adding the energy that was saved the hour before and the energy that is utilized right now to subtract the energy that was used to provide the load. Equation (6.30) establishes the accessible energy, while Eq. (6.31) sets a limit. $E_{m,h}^{\text{bat}}$ is the State of Charge (SOC) of battery, and $\eta_{\text{charge}}^{\text{bat}}$ is the charging efficiency of battery storage. $E_{\text{max}}^{\text{bat}}$ and $E_{\text{min}}^{\text{bat}}$ also represent the maximum and minimum capacity of battery storage that the SOC of battery should be within their range. δ shows a small value for self-discharging consideration of the battery.

$$E_{m,h}^{\text{bat}} = (1-\delta) \cdot E_{m,h-1}^{\text{bat}} + \left(E_{m,h,\text{charge}}^{\text{grid}} + E_{m,h,\text{charge}}^{\text{fc}} E_{m,h,\text{charge}}^{\text{pv}}\right) \cdot \eta_{\text{charge}}^{\text{bat}} - \left(E_{m,h,\text{discharge}}^{\text{bat}} / \eta_{\text{discharge}}^{\text{bat}}\right) \tag{6.30}$$

$$E_{\text{min}}^{\text{bat}} \leqslant \& \, E_{m,h}^{\text{bat}} \leqslant E_{\text{max}}^{\text{bat}} \tag{6.31}$$

The energy stored in batteries is released to power a load. In Eq. (6.32), the energy discharged must not be more than the battery's ability to discharge. $u_{m,h,\text{discharge}}^{\text{bat}}$ is a binary value that if the battery is in discharging mode, its value is equal to 1; otherwise, it is equal to zero.

$$E_{m,h,\text{discharge}}^{\text{bat}} \leqslant N \cdot u_{m,h,\text{discharge}}^{\text{bat}} \tag{6.32}$$

As stated in Eq. (6.33), the storage cannot charge and discharge simultaneously.

$$u_{m,h,\text{charge}}^{\text{bat}} + u_{m,h,\text{discharge}}^{\text{bat}} \leqslant 1 \tag{6.33}$$

6.2.2.5 Demand Response Program (DRP)

This study examines the HES's cost-emission operation while considering DRP. Within DRP, there are several programs, one of which is utilized in the chapter is the time of use [23, 24]. By using DRP, a portion of the load is shifted from the costlier (peak) times to the less expensive (off-peak) periods, flattening the load profile and lowering the overall system cost [25]. Equation (6.34) provides a mathematical expression for the above phrase.

$$E_{m,h,\text{DRP}}^{\text{demand}} = E_{m,h}^{\text{demand}} + \text{TOU}_{m,h}^{\text{electrical}} \tag{6.34}$$

In Eq. (6.34), the variable load, $\mathrm{TOU}_{m,h}^{\mathrm{electrical}}$, plus the base load equals the new load resulting from the implementation of DRP. $\mathrm{TOU}_{m,h}^{\mathrm{electrical}}$ represents the updated electrical load, accounting for the time-of-use program. This variable's value may be positive or negative, indicating a reduction or increase in load. The suggested study considers a 20% increase or reduction in load, which should not exceed the base load in Eq. (6.35). $\mathrm{DRP}_{\max}^{\mathrm{electrical}}$ shows the participation of customers in DRP, and $E_{m,h}^{\mathrm{demand}}$ is the electrical load with no DRP implementation.

$$\left|\mathrm{TOU}_{m,h}^{\mathrm{electrical}}\right| \leqslant \mathrm{DRP}_{\max}^{\mathrm{electrical}} \times E_{m,h}^{\mathrm{demand}} \tag{6.35}$$

It should be highlighted that the load is moved from peak times to other times; the underlying load amount remains constant throughout the day. To put it another way, the loads that rise and fall in each day should be equal in Eq. (6.36).

$$\sum_{h=1}^{24} \mathrm{TOU}_{m,h}^{\mathrm{electrical}} = 0 \tag{6.36}$$

6.3 Input Data

The study examined a proposed on-grid HES, as shown in Fig. 6.1. The aim was to determine the impact of DRP installation on the cost-emission operation of the HES. The following graphic illustrates how the FC-PV-BAT HES and grid supplies electrical and thermal demand. Both thermal and electrical loads are the total of two distinct kinds of loads whose load and demand data are taken from [26]. The total electrical demand is the sum of the cooling demand and the requirement for power. Moreover, the thermal demand, which includes both heating and hot water needs, can be met by an FC and backup burner that charge a heat storage tank to supply the required thermal load, as illustrated in Fig. 6.1. The backup burner and FC use the purchased gas to create heat that will be stored in the heat storage tank. Table 6.1 displays the cost of the gas that was bought. Table 6.2 also provides technical information on the parts used in the thermal section, such as the backup burner, FC, and heat storage tank. The upstream grid exports electricity to power loads and recharge battery storage since it is a consistent resource. Table 6.3 displays the price of electricity purchased from the upstream grid, which varies depending on the time of year and season. It should be mentioned that PV systems may also provide electricity to the upstream grid. Another renewable resource is PV systems; the energy output is proportionate to solar radiation, which is also taken from [26]. PV systems use solar radiation to create electricity that may be used for various purposes, including powering loads, charging batteries, and selling a portion of the electricity produced to the upstream grid. It should be mentioned that Table 6.3 displays the predetermined price at which the energy is sold.

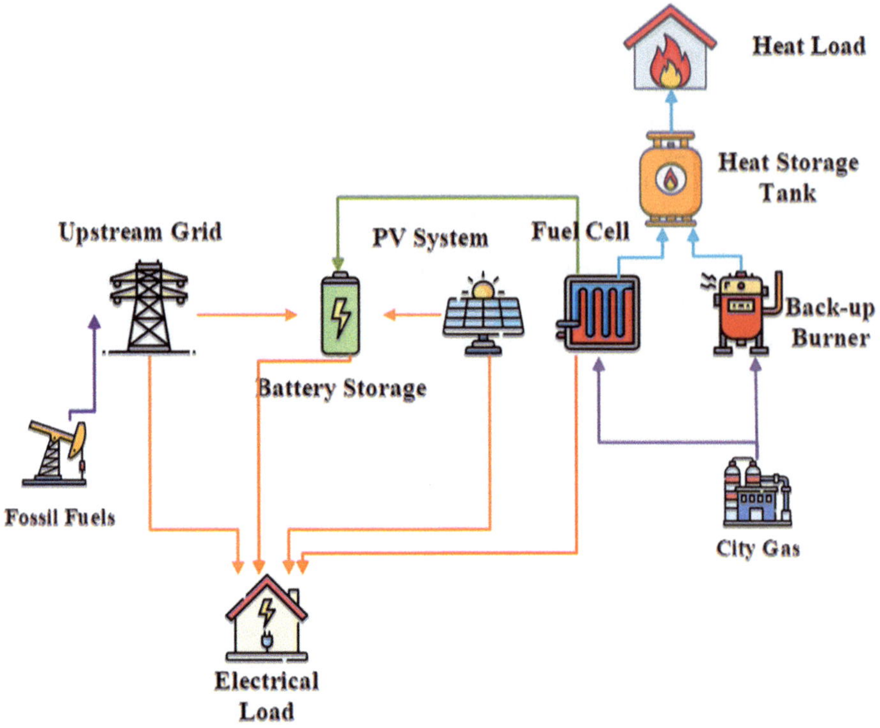

Fig. 6.1 The schematic of the studied HES

Table 6.1 Expense of gas

Section	Measurement units	Amount
Basic cost	USD/1 month	9.7
Power cost	USD/m^3	0.12

Table 6.2 The specifications of the elements used in the heating segment

Factors	Amount (%)
Thermal efficiency of a FC	0.56
Efficiency of charging heat storage tank	0.95
Efficiency of discharging heat storage tank	0.95
Heat loss in storage tank	0.05
Efficiency of backup burner	0.95

As previously stated, the FC in use can produce both heat and electricity. FCs generate electricity that is utilized for battery recharge and load delivery. Batteries, which store electrical energy from multiple sources, create an electrical demand by releasing the stored energy at varying intervals. Table 6.4 provides the necessary information regarding the FC, PV system, and battery storage that are being used. This study incorporates a year's worth of calculations and formulations. The

Table 6.3 The cost of power exchanged between the upstream grid and HES

Section	Period	Season	Measurement unit	Amount
Selling price of PV			USD/kWh	0.29
Demand charge			USD/1 month	9.7
Energy charge	Peak	Summer	USD/kWh	0.30
		Others	USD/kWh	0.25
	Mid-peak		USD/kWh	0.19
	Off-peak		USD/kWh	0.08

Table 6.4 The specifications of the elements used in the electricity segment

Parameters	Amount (%)
Efficiency of PV	0.19
Inverted efficiency	0.90
Efficiency of FC	0.39
Efficiency of charging battery	0.95
Efficiency of discharging battery	0.95
Sell-discharging value of battery	0.05

cost-emission optimization problem for the proposed HES, alongside a DRP, is addressed using the GAMS optimization software with CPLEX.

6.4 Results

This section examines the two research cases:

- Case 1: HES operating at cost-emission levels without taking DRP into account
- Case 2: HES operating at cost-emission levels using DRP into account

A PV-FC-BAT HES's cost-emission operation is examined in the first case without considering DRP. The weighted sum strategy solves the multi-objective model, yielding several solutions. All the produced solutions are ideal, but the one we're after is the one that presents a strategy that balances the interests of the two opposing goal functions. Tables 6.5 and 6.6 provide a summary of the outcomes of case 1. The optimal solution, solution number 7, with the highest $\Phi 1$ of 0.820, is chosen using the fuzzy constraint satisfaction approach. This answer indicates that the overall cost of the system is 18,265.86 and that 29,184.10 kg of CO_2 is released. These numbers may be as low as feasible with the system's components operating at peak efficiency, and the following scenario solves this issue by using DRP. The functioning of an FC-PV-BAT HES in the presence of DRP is examined in the second case regarding cost and emissions. Preceding the first instance, the weighted sum strategy is used to solve the suggested multi-objective optimization model, yielding several solutions, as shown in Tables 6.5 and 6.6. The fuzzy constraint satisfaction approach provides a mechanism for trading off conflicting goal

Table 6.5 Case 1 Pareto optimal solutions for the proposed HES

No.	(W_1, W_2)	Overall cost (USD)	Emission by CO_2 (kg)	Φ_1 (pu)	Φ_2 (pu)	min (Φ_1, Φ_2)
1	(0,1)	18,789.39	28,949.37	0	1	0
2	(0.1, 0.9)	18,497.34	28,966.32	0.457	0.988	0.457
3	(0.25, 0.75)	18,343.46	29,061.62	0.698	0.918	0.698
4	(0.275, 0.725)	18,311.62	29,085.94	0.748	0.899	0.748
5	(0.4, 0.6)	18,308.74	29,088.52	0.753	0.898	0.753
6	(0.5, 0.5)	18,274.14	29,161.49	0.807	0.844	0.807
7	(0.6, 0.4)	18,265.86	29,184.10	0.820	0.828	0.820
8	(0.75, 0.25)	18,220.4	29,465.24	0.891	0.621	0.621
9	(0.8, 0.2)	18,207.36	29,558.07	0.911	0.553	0.533
10	(0.9, 0.1)	18,152.32	30,182.79	0.998	0.094	0.094
11	(1, 0)	18,150.83	30,310.64	1	0	0

Table 6.6 Case 2 Pareto optimal solutions for the proposed HES

No.	(W_1, W_2)	Overall cost (USD)	Emission by CO_2 (kg)	Φ_1 (pu)	Φ_2 (pu)	min (Φ_1, Φ_2)
1	(0, 1)	18,487.73	28,888.14	0	1	0
2	(0.05, 0.95)	18,040.05	28,893.47	0.550	0.996	0.550
3	(0.1, 0.9)	17,910.43	28,906.08	0.628	0.986	0.628
4	(0.35, 0.65)	17,753.35	29,003.36	0.823	0.909	0.823
5	(0.6, 0.4)	17,721.51	29,027.68	0.854	0.890	0.854
6	(0.7, 0.3)	17,687.21	29,100.03	0.870	0.833	0.833
7	(0.8, 0.2)	17,665.67	29,177.69	0.894	0.772	0.772
8	(0.85, 0.15)	17,621.79	29,448.19	0.949	0.559	0.559
9	(0.9, 0.1)	17,588.75	29,758.65	0.977	0.314	0.314
10	(0.95, 0.05)	17,569.8	30,058.71	0.998	0.077	0.077
11	(1, 0)	17,567.84	30,156.90	1	0	0

functions to pick the optimal solution. Given that the highest value for the weakest membership function is 0.854, result no. 5 is the optimal choice for the selective approach. This result indicates that the overall cost of the system is 17,721.51 and that 29,027.68 kg of CO_2 is released. It is evident from examining the data that case 2's system overall cost is 2.98% less than case 1's. Furthermore, case 2's total CO_2 emissions are lower than case 1's by 156 kg, indicating a reduction in CO_2 emissions overall. The findings show that the HES's components operate at their best when DRP is implemented, which results in the best possible cost-emission functioning of the HES.

Figures 6.2, 6.3, 6.4, 6.5, 6.6, 6.7, 6.8, 6.9, 6.10, 6.11, 6.12, 6.13, 6.14, and 6.15 further explain the various components' operations in the HES for the chosen solution based on the fuzzy satisfaction method. Figures 6.2 and 6.3 show that when the DRP is implemented, the load is reduced during peak times and raised during off-peak times, flattening the load profile and lowering the overall cost of the system.

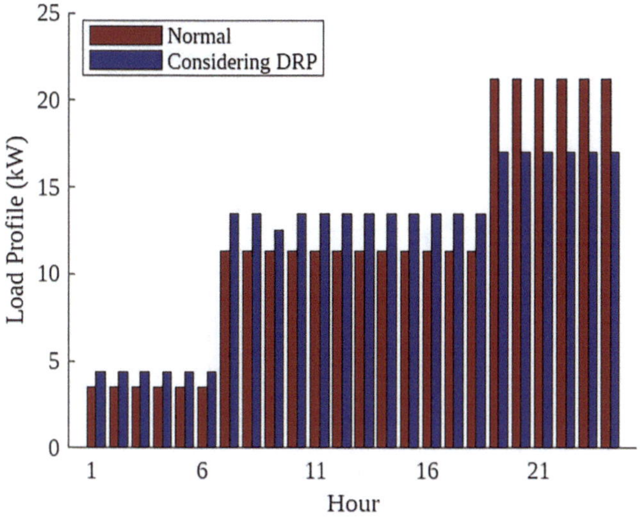

Fig. 6.2 Load profile in two cases on one day of the seventh month

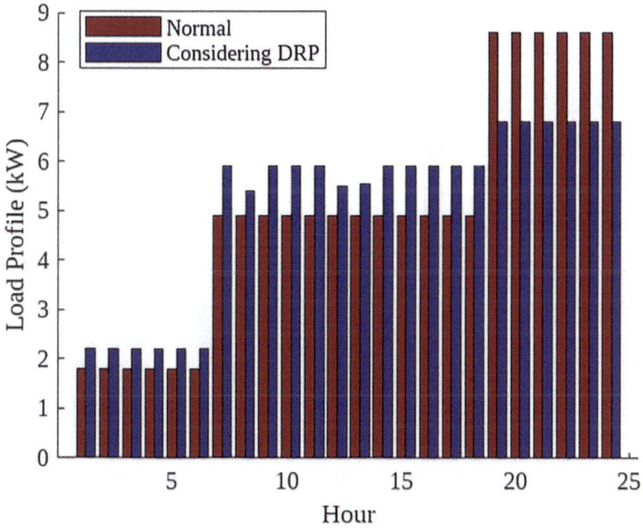

Fig. 6.3 Load profile in two cases on one day of the ninth month

Figure 6.4 shows the battery storage's charge and discharge capabilities during the first month. Figure 6.4 illustrates how using DRP reduces the charging and discharging of battery storage, extending the storage's lifespan.

As Figs. 6.5 and 6.6 demonstrate, the DRP's deployment results in less power being obtained from the upstream grid during peak hours and more energy being obtained during mid-peak and off-peak hours, resulting in a drop in overall cost.

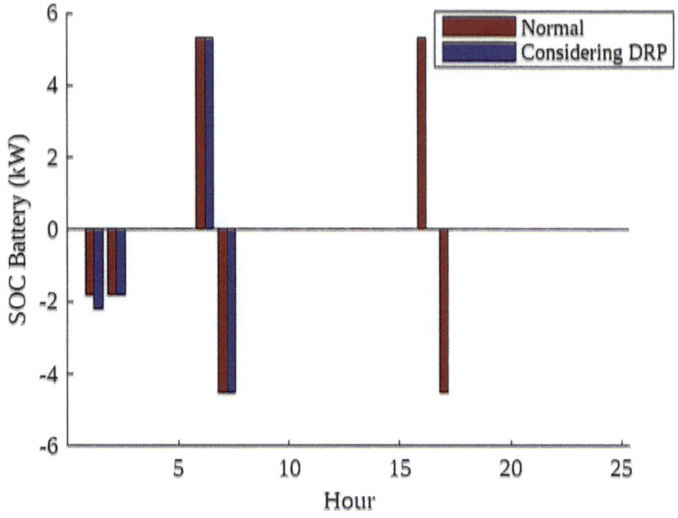

Fig. 6.4 SOC of battery in two cases on one day of the first month

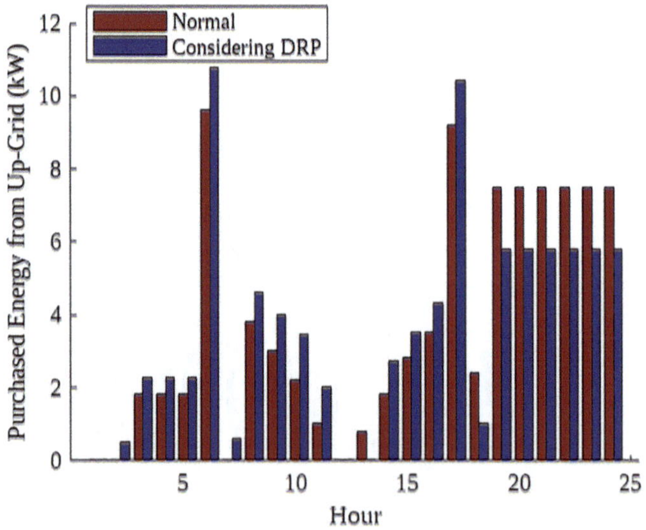

Fig. 6.5 Up-grid energy purchases in two cases on one day of the fifth month

Figure 6.7 shows the amount of PV system electricity sold to the upstream grid daily during the tenth month. This statistic indicates that all the power generated by the PV system is utilized to charge the battery and serve the load, with no extra electricity sold to lower the cost associated with power acquisition.

As seen in Fig. 6.8, the quantity of bought gas is reduced due to the use of DRP. Furthermore, as seen in Fig. 6.9, the timing of city gas purchases has been

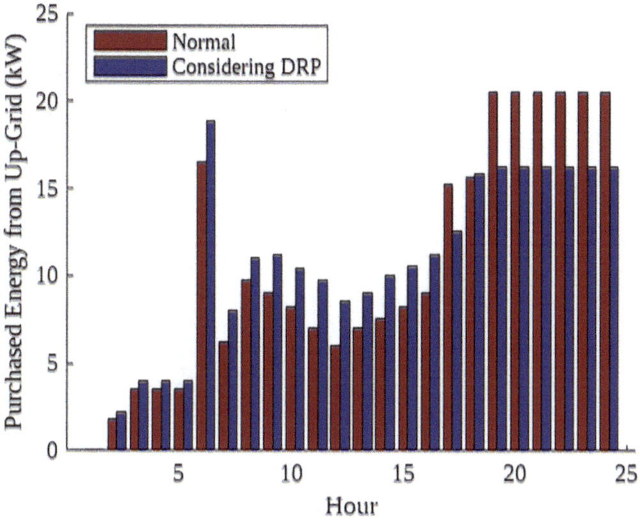

Fig. 6.6 Up-grid energy purchases in two cases in one day of seventh month

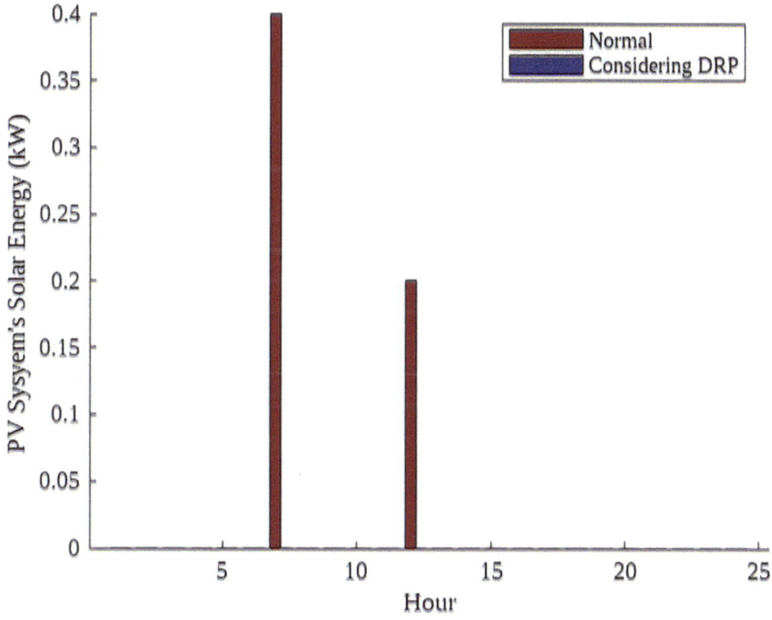

Fig. 6.7 PV system's solar energy in two cases in one day of tenth month

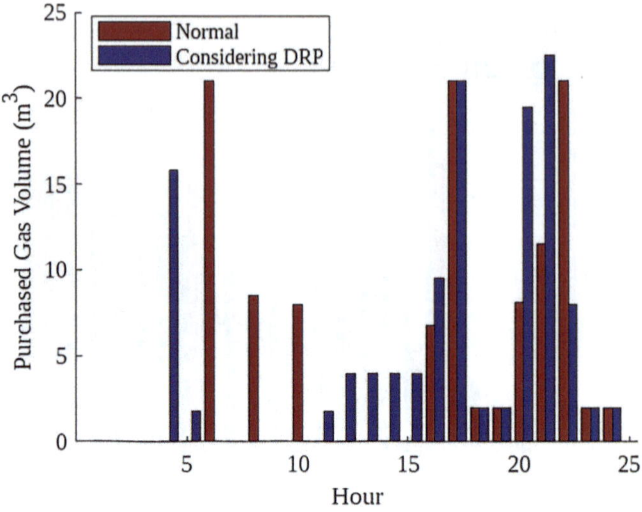

Fig. 6.8 Volume of gas purchases in one day of the fifth month

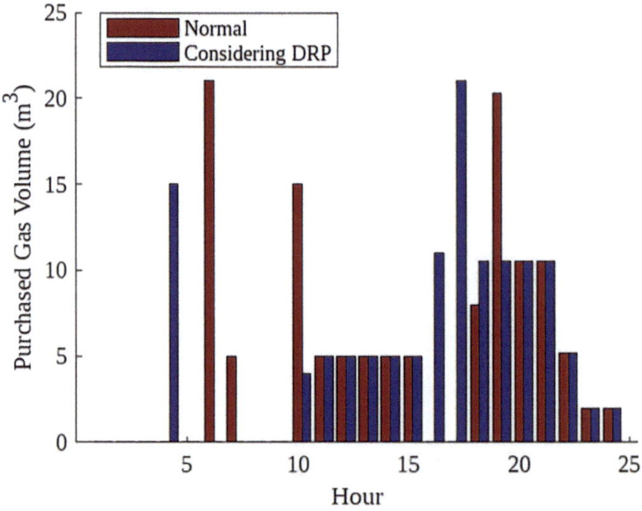

Fig. 6.9 Volume of gas purchases in one day of the ninth month

adjusted, with gas now acquired during off-peak and mid-peak hours rather than peak hours. The introduction of DRP has altered how the thermal storage system's backup burner operates as its primary heat source. Figures 6.10 and 6.11 show that the backup burner has tried to produce heat at mid- and off-peak times rather than peak times in the presence of DRP. Additionally, the backup burner produces less heat overall, extending its lifespan.

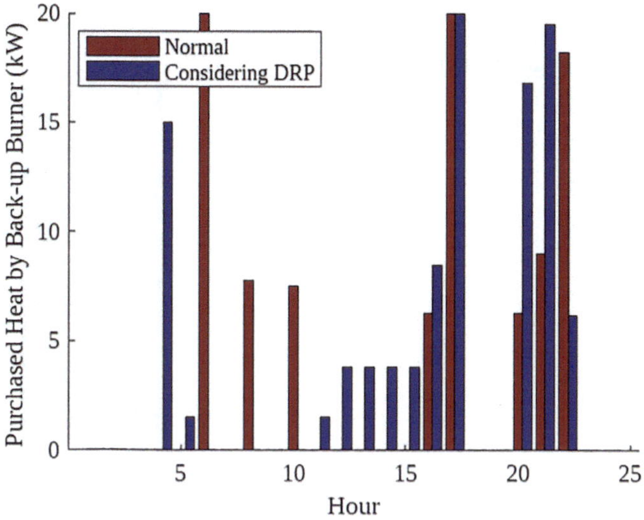

Fig. 6.10 Backup burner heat purchases in one day of the fifth month

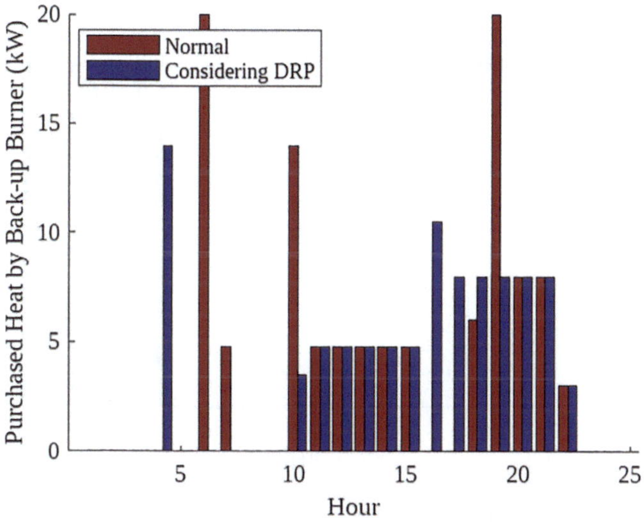

Fig. 6.11 Backup burner purchases in one day of ninth month

Figures 6.12 and 6.13 depict the heat and electricity produced by FCs, respectively. According to the statistics, FCs have been run at full capacity to produce heat and electricity during peak hours, as the heat produced by backup burners and bought energy from the upstream grid is reduced during these times. The explained figures demonstrate that the implementation of the DRP alters the generation

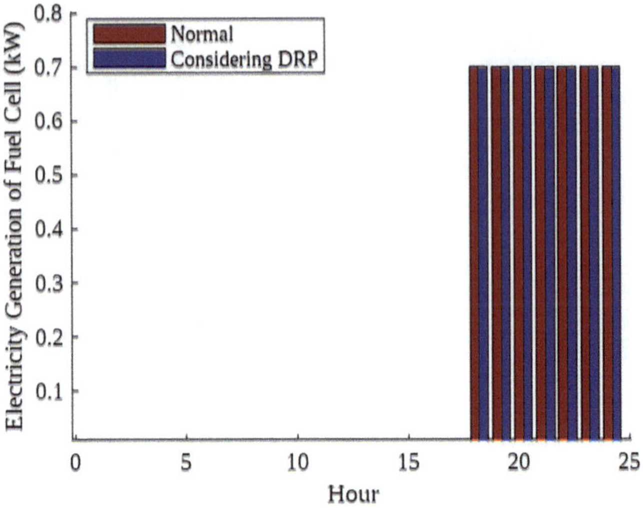

Fig. 6.12 FC's electricity generation in one day of fifth month

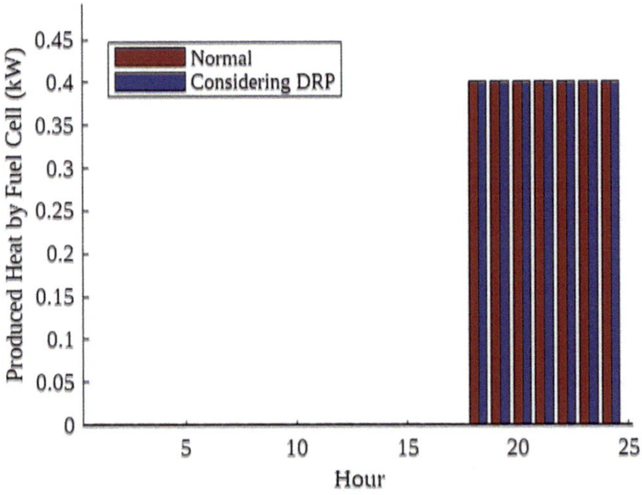

Fig. 6.13 FC's heat generation in one day of fifth month

strategy of the backup burner and FC, consequently affecting the SOC cycles of the heat storage tank. Figures 6.14 and 6.15 show that the heat storage tank tends to charge during mid- and off-peak times instead of peak times. Additionally, the frequency of SOC is reduced, thereby extending the lifespan of the heat storage tank.

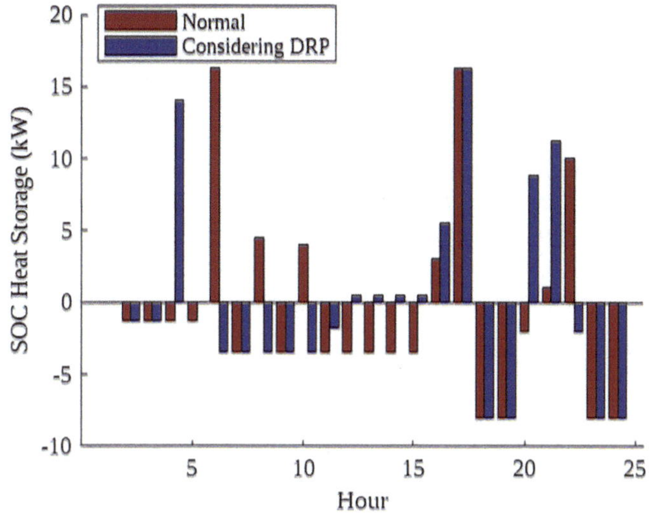

Fig. 6.14 SOC of heat storage in one day of fifth month

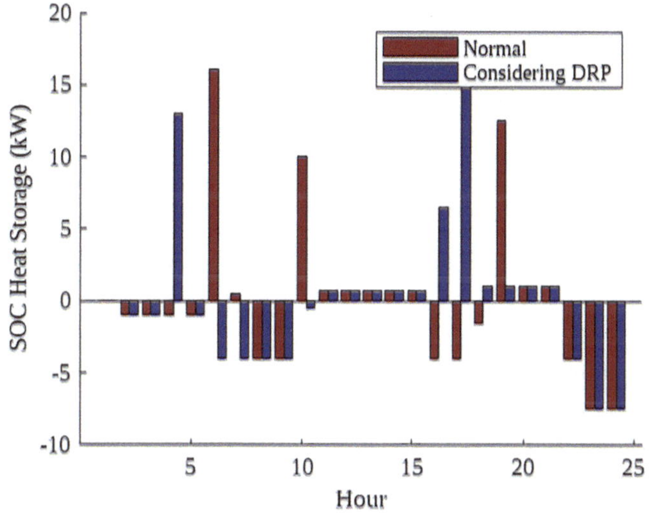

Fig. 6.15 SOC of heat storage of FC in one day of the fifth month

6.5 Conclusion

The chapter introduces a type of optimization framework designed to address the cost-emission management of a FC-PV-BAT HES, incorporating a DRP. The model is designed to simultaneously reduce costs and lower CO_2 emissions, achieving two critical objectives simultaneously. The weighted sum method addresses the suggested multi-objective framework, and a fuzzy constraint satisfaction method is applied to determine the optimal result. A sample system is examined, and two cases are simulated. The comparison of the results indicates that including the DRP leads to a reduction in the system's overall cost and a decrease in CO_2 emissions. It should be noted that implementing the DRP in case 2 will result in a reduction in operation costs. Additionally, using the DRP leads to decreased CO_2 emissions compared to the based case. Thus, the application of the DRP results in the most efficient cost-emission performance for the FC-PV-BAT HES.

References

1. I. AlHajri, A. Ahmadian, R. Alazmi, A comprehensive technical, economic, and environmental evaluation for optimal planning of renewable energy resources to supply water desalination units: Kuwait case study. Energy **275**, 127416 (2023)
2. M. Ghahramani, D. Habibi, M. Ghahramani, M. Nazari-Heris, A. Aziz, Sustainable buildings: A Comprehensive Review and Classification of Challenges and Issues, Benefits, and Future Directions, in *Natural Energy, Lighting, and Ventilation in Sustainable Buildings*, (Springer, Cham, 2023), pp. 1–28
3. M. Ghahramani, M. Abapour, B. Mohammadi-Ivatloo, M. Ghahramani, M. Nazari-Heris, Operation Optimization of Sustainable Buildings, in *Natural Energy, Lighting, and Ventilation in Sustainable Buildings*, (Springer, Cham, 2023), pp. 131–153
4. D. Akinyele, E. Olabode, A. Amole, Review of fuel cell technologies and applications for sustainable microgrid systems. Inventions **5**(3), 42 (2020)
5. C.H. Liu, J.C. Gu, M.T. Yang, A simplified LSTM neural networks for one day-ahead solar power forecasting. IEEE Access **9**, 17174–17195 (2021)
6. M. Sedghi, A. Ahmadian, M. Aliakbar-Golkar, Optimal storage planning in active distribution network considering uncertainty of wind power distributed generation. IEEE Trans. Power Syst. **31**(1), 304–316 (2015)
7. S. Sadeghi, A. Ahmadian, A. Diabat, A. Elkamel, Modeling energy management of an energy hub with hybrid energy storage systems for a smart island considering water–electricity nexus. Int. J. Hydrog. Energy **71**, 600–616 (2024)
8. M. Razeghi, A. Hajinezhad, A. Naseri, Y. Noorollahi, S.F. Moosavian, An overview of renewable energy technologies for the simultaneous production of high-performance power and heat. Future Energy **2**(2), 1–11 (2023)
9. B.K. Das, M.S.H. Tushar, R. Hassan, Techno-economic optimisation of stand-alone hybrid renewable energy systems for concurrently meeting electric and heating demand. Sustain. Cities Soc. **68**, 102763 (2021)
10. M. Daneshvar, B. Mohammadi-Ivatloo, K. Zare, Two-stage optimal robust scheduling of hybrid energy system considering the demand response programs. J. Clean. Prod. **248**, 119267 (2020)
11. M. Majidi, S. Nojavan, K. Zare, Optimal stochastic short-term thermal and electrical operation of fuel cell/photovoltaic/battery/grid hybrid energy system in the presence of demand response program. Energy Convers. Manag. **144**, 132–142 (2017)
12. M.R. Sunny, T. Ali, K. Aghaloo, K. Wang, Techno-economic feasibility of stand-alone hybrid energy system with battery storage in educational buildings: A case study of Uttara University. Energ. Buildings **304**, 113852 (2024)
13. J.S. Nirbheram, A. Mahesh, A. Bhimaraju, Techno-economic analysis of grid-connected hybrid renewable energy system adapting hybrid demand response program and novel energy management strategy. Renew. Energy **212**, 1–16 (2023)

14. R. Kaluthanthrige, A.D. Rajapakse, Demand response integrated day-ahead energy management strategy for remote off-grid hybrid renewable energy systems. Int. J. Electr. Power Energy Syst. **129**, 106731 (2021)
15. S. Basnet, K. Deschinkel, L. Le Moyne, M.C. Péra, A review on recent standalone and grid integrated hybrid renewable energy systems: System optimization and energy management strategies. Renew. Energy Focus **46**, 103–125 (2023)
16. P.T. Kapen, B.A.M. Nouadje, V. Chegnimonhan, G. Tchuen, R. Tchinda, Techno-economic feasibility of a PV/battery/fuel cell/electrolyzer/biogas hybrid system for energy and hydrogen production in the far north region of cameroon by using HOMER pro. Energ. Strat. Rev. **44**, 100988 (2022)
17. A. Emrani, Y. Achour, M.J. Sanjari, A. Berrada, Adaptive energy management strategy for optimal integration of wind/PV system with hybrid gravity/battery energy storage using forecast models. J. Energy Storage **96**, 112613 (2024)
18. Z. Abdin, W. Mérida, Hybrid energy systems for off-grid power supply and hydrogen production based on renewable energy: A techno-economic analysis. Energy Convers. Manag. **196**, 1068–1079 (2019)
19. X. Yang, Z. Leng, S. Xu, C. Yang, L. Yang, K. Liu, Y. Song, L. Zhang, Multi-objective optimal scheduling for CCHP microgrids considering peak-load reduction by augmented ε-constraint method. Renew. Energy **172**, 408–423 (2021)
20. M. Fan, Z. Li, T. Ding, L. Huang, F. Dong, Z. Ren, C. Liu, Uncertainty evaluation algorithm in power system dynamic analysis with correlated renewable energy sources. IEEE Trans. Power Syst. **36**(6), 5602–5611 (2021)
21. E.L.V. Eriksson, E.M. Gray, Optimization of renewable hybrid energy systems–A multi-objective approach. Renew. Energy **133**, 971–999 (2019)
22. M.A. Ilyas, G. Abbas, T. Alquthami, M. Awais, M.B. Rasheed, Multi-objective optimal power flow with integration of renewable energy sources using fuzzy membership function. IEEE Access **8**, 143185–143200 (2020)
23. M. Ghahramani, M. Nazari-Heris, K. Zare, B. Mohammadi-Ivatloo, A two-point estimate approach for energy management of multi-carrier energy systems incorporating demand response programs. Energy **249**, 123671 (2022)
24. M. Ghahramani, M. Nazari-Heris, K. Zare, B. Mohammadi-Ivatloo, Energy and reserve management of a smart distribution system by incorporating responsive-loads/battery/wind turbines considering uncertain parameters. Energy **183**, 205–219 (2019)
25. M. Ghahramani, M. Abapour, Optimal Energy Management of a Parking Lot in the Presence of Renewable Sources, in *2023 8th International Conference on Technology and Energy Management (ICTEM)*, (IEEE, Piscataway, 2023), pp. 1–5
26. M. Majidi, S. Nojavan, N.N. Esfetanaj, A. Najafi-Ghalelou, K. Zare, A multi-objective model for optimal operation of a battery/PV/fuel cell/grid hybrid energy system using weighted sum technique and fuzzy satisfying approach considering responsible load management. Sol. Energy **144**, 79–89 (2017)

Chapter 7
Microgrid and Nanogrid Implementation in Smart Cities

Omid Shahhoseini

7.1 Introduction

Distribution networks are transforming from passive networks to active networks, meaning that decision and control systems are distributed and power flow is bidirectional. This type of network facilitates the integration of distributed generation (DG), renewable energy sources (RES), and energy storage technologies, creating opportunities for new types of equipment and services that all operate according to common protocols and standards. The primary function of an active distribution network is to effectively link electricity generation with consumer needs and provide decision-making capabilities for optimal real-time performance. The assessment of power flow, voltage control, and protection requires cost-effective competitive technologies and new communication systems, in which information and communication technology (ICT) plays a fundamental role.

The deployment of active distribution networks requires the implementation of entirely new system concepts. Microgrids, also described as "building blocks of smart grids," may be the most promising and newest network structure. Organizing microgrids based on the control capabilities of network operation allows for the feasible integration of distributed energy resources (DERs) such as micro-turbines, fuel cells, and photovoltaic (PV) arrays, along with energy storage systems like flywheels, capacitors, batteries, and controllable loads (flexible) (e.g., electric vehicles) at the distribution level. These control capabilities enable microgrids, mainly connected to the upstream distribution network, to operate independently from the grid (islanded mode) in case of faults or other external disturbances. As a result, power quality also improves. In general, independent control is the main feature that distinguishes microgrids from distribution networks with DER.

O. Shahhoseini (✉)
Department of Power Systems Planning and Operation, Niroo Research Institute, Tehran, Iran
e-mail: oshahhoseini@nri.ac.ir

A. Ahmadian et al. (eds.), *Sustainable Energy Resources in Smart Cities*, Green Energy and Technology, https://doi.org/10.1007/978-3-031-91516-1_7

From the customer's perspective, microgrids provide thermal and electrical needs, increase local reliability, and reduce greenhouse gas emissions. By supporting voltage and reducing its drop, they improve power quality and potentially reduce energy supply costs.

From the network operator's perspective, a microgrid can be considered a controlled unit in the power system that can act as an aggregated load or generator, as well as a power source or ancillary service provider supporting the network. Therefore, a microgrid is essentially an aggregative concept with the participation of resources both on the supply and demand sides in distribution networks. Based on the local load growth and local generation, a microgrid can provide economic, technical, environmental, and social benefits for various stakeholders.

From the utility's perspective, the use of distributed generation resources can reduce the demand for distribution and transmission facilities. Naturally, the presence of resources near loads leads to a reduction in power loss in transmission and distribution circuits. Additionally, having generation close to demand can improve the quality of services provided to the final customers. Microgrids can support the network in contingencies by alleviating congestion and aiding in recovery after an error occurrence.

7.2 Definitions of Microgrids and Nanogrids

A microgrid is a group of interconnected generators, loads, and storage systems with an energy management system (EMS) to deliver local energy that precisely meets local potential needs at any given time. International agencies and various authorities have provided definitions that describe the entity and characteristics of microgrids. These definitions are as follows [1]:

IEEE Standards Association: Microgrids are localized networks that can be disconnected from the main grid to operate independently. Since microgrids can operate when the main grid is down, they can enhance grid flexibility, help reduce grid disturbances, and serve as a network resource for fast system response and recovery.

US Department of Energy (DoE): A microgrid is a collection of interconnected loads and distributed energy resources within clearly defined electrical boundaries that act as a single controllable entity concerning the grid. A microgrid can connect to and disconnect from the main grid to operate in both grid-connected and islanded modes.

Microgrid Institute: A microgrid is a small energy system capable of balancing supply and demand to maintain stable service within a specific area. There is no globally accepted minimum or maximum size for a microgrid.

CIGRE[1]: A power distribution system containing distributed loads and energy sources (such as generators, storage systems, or loads) that can operate in a controlled and coordinated method while being separated from the main grid, under the control of a unified management entity.

ASEAN Center for Energy: The microgrid system is a small power supply system composed of distributed energy sources such as renewable sources, combined heat and power (CHP) generation, fuel cells, and energy storage systems.

World Bank: Minigrid is an isolated distribution network operates on a small scale, usually under 11 kV, supplying power to a local group of customers and generating electricity from small generators. This generation potentially includes energy storage.

Lawrence Berkeley National Laboratory (LBNL): Nanogrid is a small electrical area connected to the grid with a maximum power of 100 kW and limited to a building structure or primary load or a network of loads separate from the grid, each of which does not exceed 5 kW. Both categories represent equipment capable of islanding and/or energy self-sufficiency through a level of management or control of distributed intelligent energy resources.

A microgrid is a cluster of loads and distributed resources that operates as a unified controllable system and provides electrical energy and heat for its local area. By analyzing this definition, it is easy to conclude that a microgrid is defined by its components and infrastructure. The components in a microgrid can be classified into three main categories: DERs, distributed storage systems (DS), and electric and thermal loads [2].

Microgrids consist of low-voltage distribution systems with distributed energy sources (such as micro-turbines, fuel cells, PV, etc.) along with storage systems (flywheels, capacitors, batteries, etc.) and flexible loads. Such systems can operate either independently if connected to the grid or autonomously if disconnected from the main grid. With effective management and coordination, the performance of distributed generation resources in the grid can create distinct advantages for the overall system performance [3].

It can be seen that the terms mini, micro, and nanogrid indicate the size, capacity, and configuration of both off-grid or grid-connected systems. The nanogrid and sometimes picogrid is the smallest discrete network unit capable of independent operation, such as a circuit with integrated power generation sources in a building.

7.3 Microgrid and Nanogrid Application

For a microgrid, improved reliability, cost savings, and resilience against disasters are expected goals. According to the IEC TS 62898-1 standard, the following are the scenarios for microgrids [4]:

[1] Conseil International des Grands Réseaux Électriques.

- Microgrids aimed at improving reliability and providing energy for all or part of their loads in islanded mode. Examples include distribution microgrids, such as a university campus, and facility microgrids, such as those in customer premises, military bases, and hospitals.
- Microgrids aimed at supplying electricity to remote areas with lower costs, such as isolated microgrids for rural electrification and oceanic islands.
- Microgrids aimed at reducing energy costs for microgrid users in grid-connected mode by optimizing assets such as energy storage, generators, and distributable loads and providing ancillary services to the grid.
- Microgrids that provide resilience against disasters by optimizing assets such as energy storage, generators, and dispatchable loads. These microgrids may be constructed in areas prone to natural disasters and are designed to enhance power supply for critical loads and other necessary functions.

During the planning phase, microgrids are classified as "non-isolated" or "isolated" based on the availability of a main electrical power system, local DER characteristics, and load patterns. Other determining factors include customer requirements for environmental benefits, power quality, reliability, and cost-effectiveness in microgrid. It is worth mentioning that classification should be considered as an essential stage. In this regard, the predicted goals and requirements by the network planner lead to different planning schemes. Both types of networks must comply with technical requirements.

7.3.1 Non-isolated Microgrid Application

A non-isolated microgrid is connected to the distribution system via a point of connection (POC) and can operate in grid-connected or islanded mode. Such a microgrid must be equipped with energy storage systems or dispatchable generation units. The non-isolated microgrid emphasizes the use of local resources, including renewable sources, with sufficient energy storage capacity to maintain critical load for a predetermined period in islanded mode. In a non-isolated microgrid, the energy management system (EMS) must optimize operational objectives by managing operational costs through local generation and received energy from the grid. The main distribution system should provide the required energy for critical loads with defined power quality and reliability levels. To achieve this, the non-isolated microgrid must transition seamlessly and safely from one operational mode to another while maintaining the capability to supply critical loads in islanded mode.

7.3.2 Isolated Microgrid Application

An isolated microgrid is not connected to the main distribution system and therefore operates in islanded mode permanently. Such microgrids are mainly used in remote areas far from a large-scale power system. The goal of an isolated microgrid is to provide continuous and reliable energy with or without sufficient renewable energy sources. Therefore, such a microgrid must have sufficient energy storage capacity and dispatchable DERs. An isolated microgrid must maintain power balance through generator management, energy storage system, and demand response. Planners of isolated microgrids must make an economic decision between installing large electrical energy storage systems and paying the cost for periodic fuel delivery. Scalability of isolated microgrids must be considered from the beginning through microgrid development or interconnection of microgrids.

7.4 Microgrid and Nanogrid Components

As previously mentioned, a microgrid is a group of DERs interconnected with each other and with the main grid at well-defined electrical boundaries, operating as a controllable electrical unit in relation to the main grid. The microgrid can be powered by the main grid or DERs. Therefore, essential components of the microgrid include local generation resources, DERs, loads, energy storage systems, control systems, and grid connection equipment in POC or point of common coupling (PCC) (Fig. 7.1).

The selection of a combination of microgrid components is based on the requirement for the microgrid to manage any supply–demand imbalances using intelligent management systems within an appropriate time interval. In this regard, reliability, stability, cost-effectiveness, and availability and local network features must also be considered.

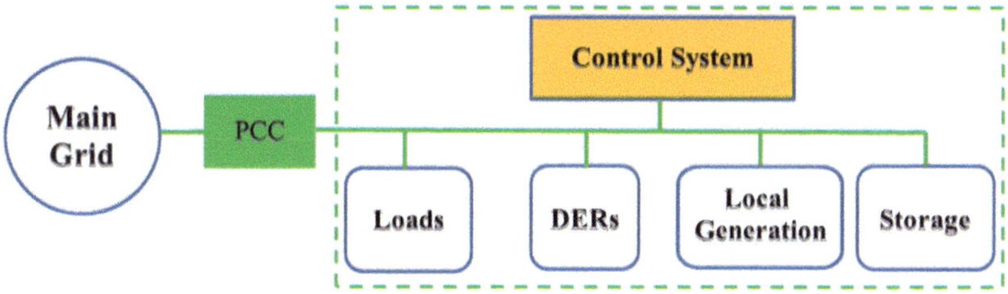

Fig. 7.1 Essential components of microgrids

7.4.1 Distributed Energy Resources

A microgrid must be able to independently supply energy to connected loads, so generation resources must exist within the microgrid. These resources can be controllable or uncontrollable and can be renewable (e.g., solar photovoltaic, wind, biomass, or less commonly used resources like solar thermal, micro-hydro, and tidal) or nonrenewable (e.g., diesel, gas turbines, internal combustion engines, and micro-turbines). The choice of energy source should be based on the needs and constraints of the microgrid. Determining the required generation capacity based on available renewable and fuel sources, along with storage requirements, is crucial. The existing generation resources affect the distributed generation capacity added to support the microgrid. The reduction in the cost of solar photovoltaic modules and other types of renewable energy, along with energy storage, has increased the economic feasibility and attractiveness of renewable energy-based microgrids.

7.4.2 Loads

Final energy demand determines the electrical loads that the microgrid must be able to supply. These loads in turn impact the generation capacity and storage needs. For example, charging a mobile phone imposes a minimum capacity requirement on the system, while connecting an air conditioning system to a solar PV-based microgrid can create challenges in managing electrical loads due to high continuous energy consumption. In reality, a wide spectrum of loads must be met by the microgrid.

In terms of controllability characteristics, microgrid loads vary from critical loads such as data centers or hospital backup equipment at one end of the spectrum to adjustable loads such as air conditioning and lighting at the other end. The second category of loads can be temporarily reduced or adjusted flexibly according to consumer needs. Most modern microgrids have the capability to control these loads to optimize the use of generation resources and storage systems. Any type or class of supply, demand, and even storage can be classified based on controllability, reliability, and criticality.

7.4.3 Energy Storage Systems

Microgrids must have an energy storage system to maintain their generation at different times, in order to meet the demand and also to serve as a vital source of generation in islanded mode. Through energy storage, microgrids can absorb and store the generated energy when supply exceeds demand and return it when demand exceeds supply (e.g., during peak evening hours when solar energy generation is not available). While there are some inefficiencies in energy storage, those are minimal compared to the value that it can provide.

Energy storage technologies are chosen based on the size of the microgrid and the required response time. For remote or smaller microgrids connected to the grid, batteries are the most common storage technology because the storage capacity required for these microgrids does not justify the higher costs associated with other storage technologies. Batteries can also be used to provide ancillary services needed in microgrids. Large-scale storage technologies, such as pumped hydro or thermal storage systems, although cheaper for time-shifting operations, have high initial costs that make their implementation challenging for smaller networks and microgrids.

7.4.4 Monitoring and Control Systems

Microgrid Monitoring and Control Systems (MMCS) connect all microgrid components and maintain the balance of generation and consumption in real time. These control systems can have a centralized or decentralized design. In the simplest microgrid configuration, the control system may only consist of a governor that controls a diesel generator. In more complex microgrids, MMCS includes sensors, hardware, middleware, complex platforms, and communication systems designed and implemented to optimize and control generators, energy storage systems, loads, and user interfaces in real time. When microgrid is connected to the grid, MMCS must manage grid connection and communication with the system operation center or independent system operator. In grid-connected mode, depending on demand and generation, a microgrid either receives energy from the main grid or injects energy into the grid. During this state, the static switch is "closed." All critical and noncritical loads provide their power from the primary grid. Distributed generators in microgrid operate in parallel with the primary grid to meet the power requirements of these loads.

7.4.5 Power Electronic Converters

Power electronic converters are used for connecting and managing various components in a microgrid. For example, DC loads need converters to be connected to AC generators. To match the input and output voltages of a microgrid element, various conversion stages such as DC-AC, AC-DC, or DC-DC may be required. DERs and loads are generally connected to microgrid distribution lines through power electronic converters (Fig. 7.2). They adapt microgrid current and voltage levels to connected units. Power electronic converters vary based on the type and parameters of the microgrid, but generally, power conversion efficiency in DC systems is higher

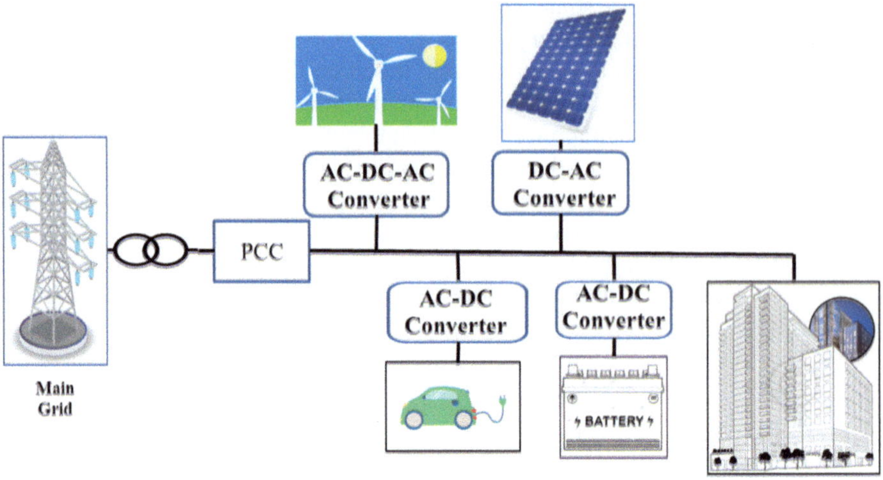

Fig. 7.2 Power electronic converters of a typical microgrids

than in AC systems and primarily depends on the primary source technology and the AC/DC load ratio.

7.4.6 Protection Devices

Protection is a crucial principle in the design of microgrids and nanogrids. The protection system must act almost instantly and be carefully planned to distinguish between grid-connected and islanded operating states. In the grid-connected mode, protection is simpler due to the potential for large fault currents, whereas in the islanded mode, the fault currents may be relatively small due to the existence of power electronic converters. This low fault current in the islanded mode is not sufficient for conventional protective relay tripping. Therefore, an adaptive protection system is required to adjust real-time relay settings to ensure continuous protection of the microgrid and nanogrid. Another solution could be the use of networked digital relays for protection. An easier approach to addressing the protection issue is to design the microgrid and nanogrid in a way that, in the event of a fault, it transitions to the islanded mode before any protective action is taken.

7.4.7 Communication Systems

Modern communication systems are essential for ensuring the stable, reliable, and optimal performance of a microgrid. The presence of DERs and loads, as well as the interaction between all points in a microgrid, significantly complicates the

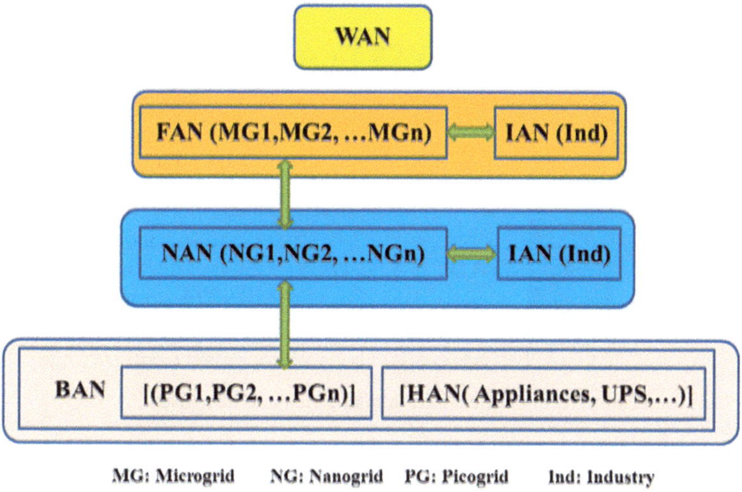

Fig. 7.3 High-level configuration of the microgrid and its communication networks

operation and control of the system and consequently demands robust communication. The communication network of a microgrid requires a high level of reliability, with bidirectional and mutual connection between microgrid resources. In a way, the communication network of a microgrid can be considered as a bridge between its physical infrastructure and the control and protection processes. This is in contrast to the traditional power network structure, which usually lacks the aforementioned connection, as the system operates as a unidirectional centralized system. The tools, standards, and architecture of communication systems determine how the components of a microgrid communicate with each other using one or more protocols, through wired (fiber optic, copper) or wireless (radio, cellular) routes.

Microgrids connected to the power network are rarely independent, but rather interact and communicate on one side with the traditional power network and on the other side with industrial, commercial, organizational, residential, and other consumers that may have subnetworks. As shown in Fig. 7.3, microgrids and the networks they interact with are generally classified into WAN,[2] IAN,[3] FAN,[4] NAN,[5] BAN,[6] and HAN[7] networks [5]. The communication technologies and standards used in them are illustrated in Fig. 7.4.

[2] Wide area network.

[3] Industry area network.

[4] Field area network.

[5] Neighborhood area network.

[6] Body area network.

[7] Home area network.

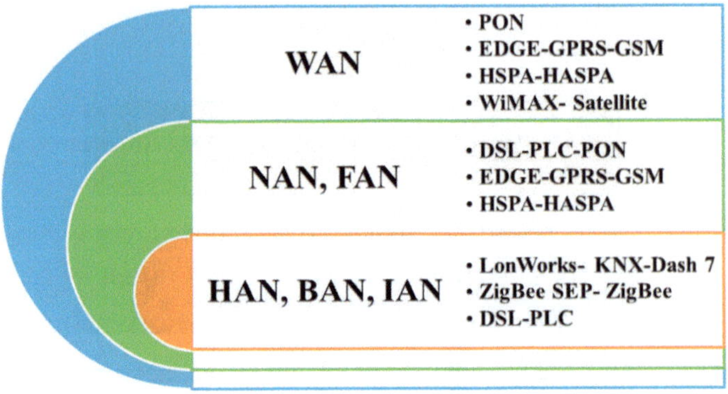

Fig. 7.4 Communication technologies used in microgrids and nanogrids

To ensure reliable and secure operation of communication networks, two essential requirements must be observed. The first requirement is real-time performance and security maintenance through confidentiality, integrity, and availability of information, and the second requirement is real-time boundary delineation of the microgrids. These requirements are challenging because most digital systems in operation, including sensors, control subsystems, and control logic execution, introduce delays in the communication networks. Additionally, many communications devices lack hardware support for encryption algorithms and therefore cannot utilize expensive computational security algorithms.

7.5 Microgrid and Nanogrid Design

7.5.1 Implementation Stages of Microgrid and Nanogrid Projects

From the network management perspective, assessment, design, construction, and commissioning of a microgrid or nanogrid follow the same extensive stages pursued by other infrastructure projects with similar stakeholders. For existing facilities or the Brownfield microgrid, pre-project assessments begin with an evaluation of the current status. In both the Brownfield and Greenfield projects, then a high-level assessment is conducted comprising the following stages [5]:

- Technical assessment including current and historical levels of reliability of power sources, current generation mix, type of distribution network, typical load characteristics, identification of critical loads versus adjustable loads, heating and cooling loads (if applicable), and available space for the microgrid.
- Environmental assessment including emission rates and emission targets.

- Financial assessment including operational costs, fuel costs, electricity prices, fuel price, fluctuations, the ratio of electricity costs as a share of the total cost of commodities sold, and certain costs arising from reliability.
- Project objectives including minimizing energy bills, reducing electricity outages, reducing greenhouse gas emissions, ensuring spinning reserves, reducing peak loads, and so on.

After the assessment, it's time to move on to the feasibility study. Project feasibility study includes the following key considerations [5]:

- Implementable policies including planning and permit regulations, power tariff structure, network connection costs, and network usage costs.
- Renewable resources including wind speed, solar radiation, shading effects, etc.
- Location information including land availability, planning permit thresholds, environmental impacts, visual impact assessment, development scheduling, and supply chain.
- Business structures including private ownership, infrastructure capital, Independent Power Producer (IPP), Power Purchase Agreement (PPA), and return to owners.
- Financial structure including balance sheet, shareholder debt/profit, Public–Private Partnership (PPP), etc.
- Technical sustainability including network integration principles, optimal technology and size selection, site design, and power return capability to the grid.

The next step is the investment evaluation. A summary of social, economic, technical, legal, and regulatory criteria that must be considered as part of the microgrid investment project evaluation has been provided in Table 7.1 [5].

The final step is Engineering, Procurement, and Construction (EPC) stage includes precision engineering, procurement, installation and assembly, commissioning, operations, and maintenance.

7.5.2 Design Parameters

Once the feasibility study was completed and the project received the green signal, it's time for the design phase. The aspects that need to be considered conceptually and precisely in the design of the microgrid system are listed below. The relevant process flow is shown in Fig. 7.5 [5]:

- Demand forecasting.
  - Number of customers.
  - Circuit lengths.
  - Load analysis (such as criticality, controllability, high efficiency, etc.): Dealing with low-income customers with limited financial and technical

Table 7.1 Social, economic, technical, legal, and regulatory evaluation criteria for microgrids

Class	Criterion	Description
Social	Management	Facilitating the management of the power distribution system
	Justice	Equality in the amount of electricity supplied to each family
	Household benefits	Improving the quality of life and income of families
	Social services	Electricity supply to social services (schools, health centers, mosques)
	Production activities	Creating local businesses through the development of production activities
	Impact on local resources	Land covered with the installed equipment
	Clean energy	Renewable energies with variable or hybrid generation
Economic	Investment	Initial investment, operation, and maintenance costs
	Tariff	Consumers' efforts to pay electricity tariffs
	Subsidies	Financial assistance or support provided by the government or governmental/private authorities
	Fuel efficiency	Reduction of losses, heat, cooling, power, etc
	Ancillary services	Relief of congestion, frequency regulation, reactive power control, and voltage control, etc
Technical	Application	Industrial, organizational, military, residential, remote, or rural areas
	Supplied energy	The amount of energy and power supplied to each family
	Resource sustainability	Reliability of consumed energy resources and battery independence
	Flexibility	Expansion of users and increased consumption
	Local substitution	Assistance in repairing equipment failures
	Local generation	Proximity to locations where equipment is manufactured
	Energy security	System reliability in case of failure equipment, cascading outages, cyber and physical attacks
Legal and regulatory	Regulatory issues	Government or private regulations
	Connection	Regulations for small power generators, load aggregators
	Electric company regulations	Regulations by governmental or private authorities

resources is somewhat challenging, so it may be necessary to use estimation and extrapolation.

- Generation capacity.
  - Generation mix.
  - Capacity and type of renewable resources (optimal capacity of photovoltaic and wind–solar systems).
  - Capacity of nonrenewable generation.

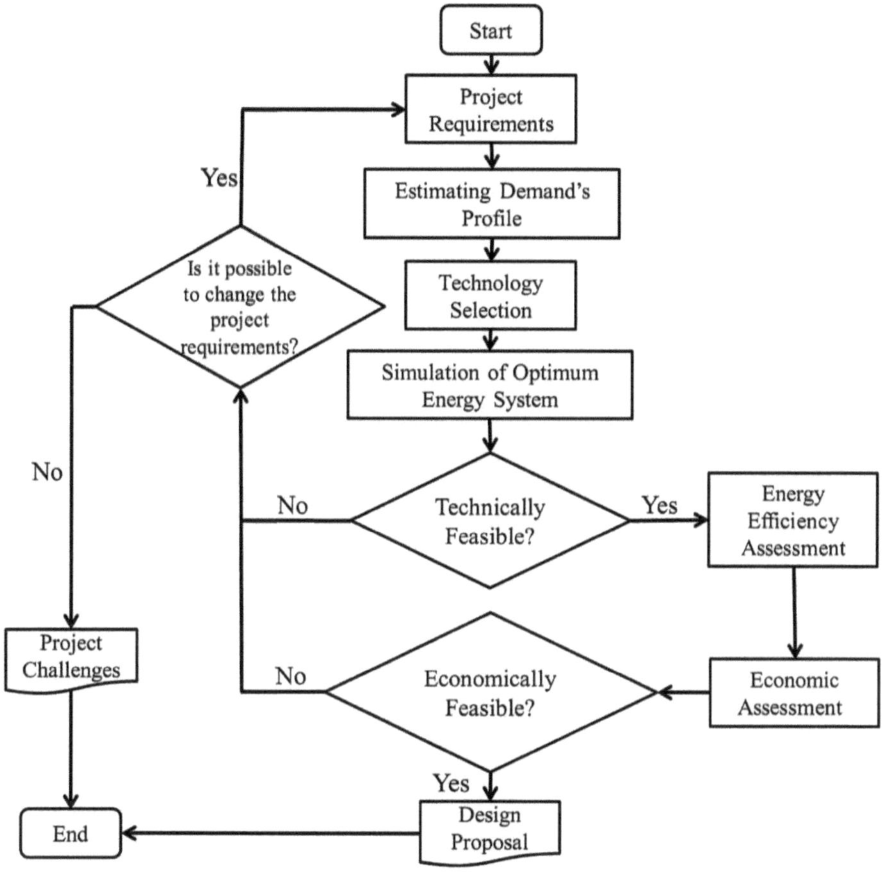

Fig. 7.5 Microgrid design algorithm

- Selection, combination, and capacity of energy storage equipment.
  - Power losses in standby mode (when the microgrid is islanded, it requires storage. Power losses in standby mode reduce the efficiency of the microgrid).
  - Response time (for seamless transition, the response time must be very fast, more than battery response time, as communication delays and control functions also play a role).
- Technical design.
  - Microgrid controller architecture and design (centralized vs. decentralized).
  - Type of power technology (AC, DC, or hybrid microgrid).
  - Voltage levels.
  - Feeder structure.
  - Desired power quality and reliability levels.

- Protection method—several protection schemes have been developed for centralized and decentralized AC microgrids. For DC microgrids, this area is less developed due to the lack of standards and guidelines, along with limited practical experience.
- Communication technology.
- System distribution modeling, simulation, and design (scenarios, network impact assessment, and model evaluation).
- Network studies (steady-state, fault analysis, protection, and stability studies).

7.5.3 Equipment Selection

Many of the equipment used in microgrids today are commercially available from several major vendors as well as several specific distributors. With increasing volume, a potential price reduction is expected and continuous improvement in performance is anticipated. Some companies also participate in the aggregation sector.

7.5.4 Capital Costs

The total capital costs of a microgrid include expenses arising from its initial development stages, including feasibility studies, permits, design, land acquisition, EPC, and commissioning. These costs may vary depending on the application, level of reliability, complexity, whether it is a Greenfield site without existing equipment or a Brownfield site with existing assets, and, to some extent, the geographical region. However, it seems that the highest cost is related to the distributed generation equipment costs, followed by the microgrid control system.

7.5.5 Simulation and Design Software

Various tools exist for simulating and analyzing microgrids. The use of computer simulations provides a simpler understanding of the phenomena and concepts surrounding these systems. GridLAB-D, developed by the DoE through PNNL,[8] is an open-source software that offers a flexible and agent-based modeling approach. It processes input files (XLM and GLM) and generates simulation results in output files. GridLAB-D effectively enables the simulation of large systems and allows the visualization of modeled objects representing various elements present in distribution systems.

---

[8] Pacific Northwest National Laboratory.

HOMER Energy, established by NREL[9] in 2009, offers HOMER PRO and HOMER GRID, both of which perform technical optimization to find the least expensive system that meets the imposed technical constraints. MATLAB, with its MATLAB Simulink platform, includes MATLAB Simscape Electrical as a solution for simulating electronic components, mechatronics, and power systems. This tool is commonly known as SimPowerSystems, and many publications still refer to it by this name.

OpenDSS is an open-source software for simulating power distribution systems. It has been available since 1997 and has been used by EPRI[10] to facilitate studies on connecting DERs in the field of smart grids.

PSCAD is a tool for simulating electromagnetic transient states. Developed in 1976, this program has a user-friendly interface widely used for electromagnetic transient analysis. The visual interface provided by PSCAD supports circuit modeling, real-time control, decomposition and analysis, and report generation.

The increasing importance of microgrids has led to a demand for study tools and complex HIL[11] testing capabilities to evaluate the performance of control and power equipment in the network. The major advantage of RTS[12] platforms is their ability to communicate with external equipment, commonly referred to as Devices under Test (DUT). In this context, three real-time simulation platforms have been developed:

1. OPAL-RT.
2. Typhoon HIL.
3. RTDS.

OPAL-RT Technologies, founded in 1997, is a Canadian company that has introduced RTS to meet the needs of industry and research institutions. The OPAL-RT simulation platform uses two main platforms including HYPERSIM and RT_LAB. To better understand the differences between simulation platforms, OPAL-RT has provided a comparative table in [6] and a list in [7]. OPAL-BROCHURE [7] organizes relevant information and provides specific applications for each platform. Additionally, a comparative chart is presented based on the frequency of phenomena and the number of nodes in the simulated system in [7].

Typhoon HIL is a company established in North America in 2008, offering three RTS platform series consisting of 4, 6, and M. The Series 4, HIL402 model, is the size of a notebook and uses software called HIL Control Center. The Series 6 focuses more on centralized power electronic equipment. The HIL602, with a maximum of six cores, weighs approximately 4 kg. The Series M is specifically used for

[9] National Renewable Energy Laboratory.

[10] Electric Power Research Institute.

[11] Hardware in the loop.

[12] Real-time simulator.

testing controllers and comes in several sizes. This platform includes schematic editor modules (for circuit modeling), HIL SCADA (which allows users to reconfigure the test setup), a test suite (comprising a test program and certification issuance), and power system toolbox (specifically developed for microgrid analysis) [8].

RTDS refers to the RTS platform produced by the Canadian company RTDS Technologies. This company was responsible for introducing the first all-digital power system simulator in 1989, becoming a global reference for real-time power system simulation. The main applications of RTDS are in the fields of microgrid/smart grid simulation, cyber security,[13] PHIL,[14] HVDC,[15] FACTS, protection system testing, control system testing, power electronics,[16] PMU testing, IEC 61850, and DER aggregation. In grid analysis with RTDS, physical equipment operates under similar conditions to real-world applications. An exclusive software package called RSCAD is used to build models. This package has been developed to RTDS in order to provide a complete graphical interface facilitating the high-speed simulation platform's operation [9].

7.6 Microgrid and Nanogrid Standards

Microgrids exist in various locations with diverse applications. Various technical standards accepted globally establish the laws necessary for planning, designing, and developing microgrids. Such recommendations lead to reducing system deficiencies. IEEE and IEC standards help determine the distinctive features of a microgrid and explain its real-time performance. There are multiple standards for microgrids, with this chapter focusing on the most important ones.

7.6.1 IEC TS 62898-1

This standard is a section of IEC 62898 including technical specifications for the design, planning, and implementation of microgrid projects. The microgrids considered in this standard consist of AC electrical systems with DERs at low or medium voltage levels. This standard does not cover DC microgrids [4].

[13] Power hardware-in-the-loop.

[14] High voltage direct current.

[15] Flexible alternate current transmission systems.

[16] Phasor measurement unit.

7.6.2 IEC TS 62898-2

The aim of this standard is to provide guidelines for the operation of microgrids. The microgrids considered in this standard are AC electrical systems with distributed energy resources at low or medium voltage levels. This standard does not cover DC microgrids [10].

7.6.3 IEC TS 62898-3-1

The purpose of this section of the IEC 62898 standard is to provide technical guidance for dynamic control and fault protection in microgrids. Dynamic control and protection in a microgrid are aimed at ensuring the safe and stable operation of the microgrid under fault and disturbance conditions [11].

7.6.4 IEEE 2030.7

This standard provides the requirements and technical specifications for microgrid controllers. It also includes attachments containing the description of the microgrid, operational specifications, the structure of microgrid control functions, and bibliography. These attachments are for informational purposes only and can be referenced, but their use with this standard is not necessary [12].

7.6.5 IEEE 2030.8

This standard is a set of test processes that enables the examination, quantifying the performance, and comparison of performance with minimum expected requirements for various functions of the microgrid controllers. Its goal is to provide criteria for comparing the control functions required by the microgrid operator and the distribution system operator [13].

7.6.6 IEEE 2030.9

This standard focuses on parameters that must be considered for the planning and design of microgrids. It provides good approaches and methods in microgrid planning and design, including system configuration, electrical system design, safety, monitoring, power quality control, electric power measurement, and project evaluation. This standard applies to AC microgrids that can be connected to the grid or operate independently [14].

7.6.7 IEEE 2030.10

This standard describes the architecture of a DC microgrid for rural and remote applications with a nominal distribution voltage of 48 V. It also defines voltage and power quality standards for electricity supplied to loads connected to such a microgrid. This standard covers the power distribution section of a microgrid. It does not impose minimum or maximum current limitations [15].

7.6.8 IEEE 2030.12

This standard is currently in draft form and is being finalized and published in the IEEE forum. This standard covers the design and selection of protective devices and their coordination for various microgrid operating modes, including grid-connected and islanded modes, and transitions between modes [16].

7.6.9 IEEE 1547

The main focus of this standard is the requirements, technical specifications, and tests for the connection of DERs to the distribution network. These requirements pertain to performance, operation, tests, safety considerations, and maintenance of the connection. They also include general requirements, response to abnormal conditions, power quality, islanding, and the necessary test requirements for design, production, evaluation, installation, commissioning, and periodic testing. The requirements for connecting distributed energy resources include synchronous machines, asynchronous machines, or inverter-based resources. These criteria and requirements are applicable to all distributed generation technologies connected to the distribution network. The primary emphasis of this standard is the installation of these resources in radial distribution networks, although it encompasses all distribution networks. This standard considers a frequency of 60 Hz for the resources [17].

7.6.10 IEEE 1547.1

Communication equipment connecting DERs to a power system must comply with the requirements of IEEE 1547. Standardized test procedures are necessary to establish and confirm compliance with these requirements. This standard includes test procedures for validating the connection of DER equipment to the power system in accordance with the IEEE 1547 reference standard. These procedures include type tests, production tests, and commissioning. Evaluation of installations and periodic testing are required in accordance with the IEEE 1547 standard [18].

7.6.11 IEEE 1547.2

The IEEE 1547 standard guide is a technical guide that addresses the operational requirements. This document provides technical explanations, schematics, practical guides, and examples of connecting energy sources for easier use of the IEEE 1547 reference standard after defining the generalities about the connection system, DERs, and local and regional power grids [19].

7.6.12 IEEE 1547.3

This standard is a guide for monitoring, data exchange, and control of the connection between DERs and power systems. It includes control and data exchange, data monitoring, data exchange models, protocols, communication infrastructure, data communication, and security issues including cyber security [20].

7.6.13 IEEE 1547.4

This standard is a guide that includes approaches for the design, operation, and integration of DERs in islanded and grid-connected modes. These include the ability to isolate from the local power grid and reconnect to it [21].

7.6.14 IEEE 1547.6

This standard addresses the connection of DERs to the secondary distribution network based on the IEEE 1547 standard. It specifies the criteria, requirements, and tests related to this type of connection [22].

7.6.15 IEEE 1547.7

This standard provides criteria for conducting engineering studies on the impact of DERs connected to the grid on the distribution system. The study scope is defined as a function of the discernible characteristics of DERs, the electric power grid of the region, and the connection. General considerations include studies on the impact of DERs on the regional power system, primary energy source considerations, specific technology considerations of the energy source, and considerations related to the regional power system. Considerations for DERs affecting the connection include the type of energy source, whether it is an inverter or non-inverter supply, the relevant protection system, communication and monitoring system, control system, operating modes and connection conditions, disconnection, and reconnection. Considerations for the regional power system affecting the connection include system arrangement, grounding type, relevant protection system, communication and monitoring system, control system, system strength at the point of common coupling, and automatic recovery system. Additionally, this standard provides examples of these cases [23].

7.6.16 IEC 62257 Series

The first part of this series introduces a method for implementing rural electrification using standalone renewable energy systems independent of the grid. It also provides guidance for facilitating the reading and use of the 62,257 series standards for decentralized rural electrification in developing or developed countries. Other parts of this series are as follows:

1. Parts 2–6 are supportive of the cognitive method for project management and implementation.
2. Parts 7–12 cover technical specifications for individual or aggregated systems and related components.

7.7 Microgrid and Nanogrid Projects in Developed Countries

Microgrids are generally established for experimental purposes, although in recent years some have entered the commercialization phase through pilot projects. Nevertheless, expanding some of the adopted solutions in microgrids (such as electrical storage systems) for larger applications is challenging. This section discusses examples of implemented microgrids in developed countries. The focus of this section is on microgrids that are larger than laboratory-scale microgrids and can represent distributed generation in a region.

7.7.1 The USA

In the United States, in order to reduce peak load, a large number of research projects have focused on the development of microgrids. Among them, the implementation project of microgrid by CERTS is notable, which is a full-scale test bed installed near Columbus, Ohio. This microgrid is operated and managed by American Electric Power. Schematic of the CERTS Microgrid has been shown in Fig. 7.6.

There are different feeders, some of which can be isolated from the network using a static switch, while the loads can be remotely controlled. The CERTS team has mainly focused on the electrical characteristics of each source, although a set of thermal characteristics are equally considered. Among the sources, there can be a directly connected machine (usually consisting of synchronous or induction

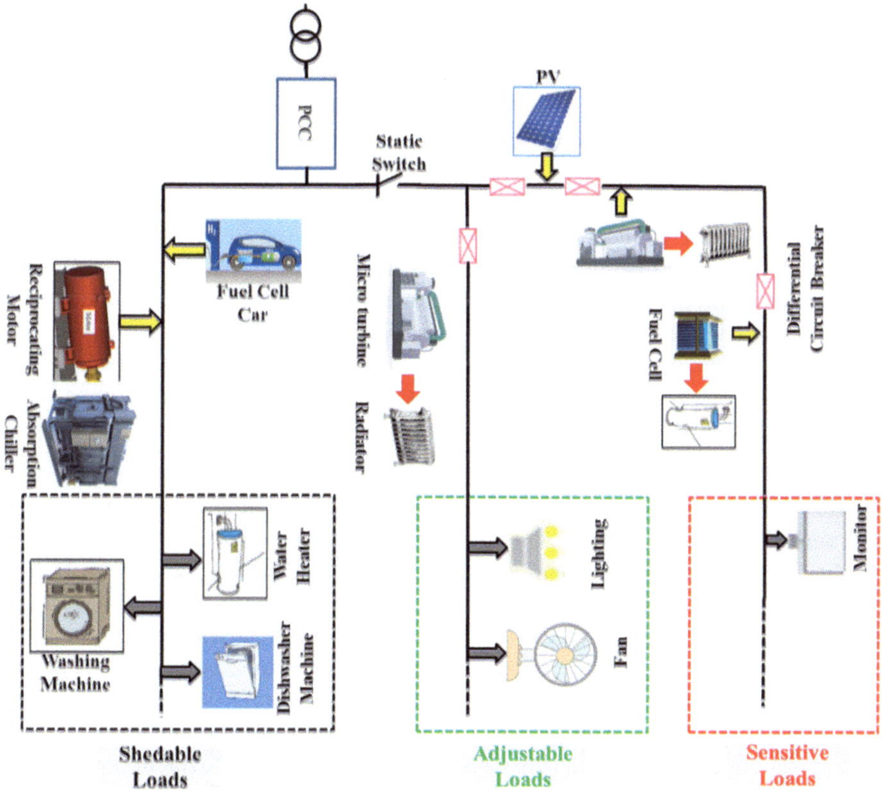

Fig. 7.6 Schematic of the CERTS microgrid

machines connected to a primary mover such as a motor or turbine) and a machine connected with an inverter. Each source can have several primary movers as inputs, such as solar energy, wind, energy storage, fuel cells, motor, and turbine. The load shedding process can be done in a centralized manner by an energy management system or in a distributed manner with predefined load shedding criteria, or a combination of the two systems [24].

In relation to the CERTS project, two software have been developed. The first software is mu-Grid, developed by the Georgia Institute of Technology. The second item is the Distributed Energy Resources-Customer Adoption Model (DER-CAM), used at the LBNL [25].

The CERTS project is part of the Renewable and Distributed Systems Integration (RDSI) program, managed by the United States Department of Energy. This program includes several implementation projects in the United States involving technologies such as photovoltaic, fuel cells, hydro turbines, energy storage systems, plug-in hybrid vehicles, wind turbines, compressed air energy storage, and automated demand response systems. For example:

- The Never-Failing Perfect Power Prototype, developed by the Illinois Institute of Technology (IIT), includes distributed generation, CHP, and renewable energy. The IIT campus peak load is about 10 MW. DER inside the campus includes two 4 MW combined cycle power plant units and a small wind turbine, with plans to add photovoltaic systems on rooftops, as well as a 500 kWh battery. The total DER capacity will be close to 9 MW. Therefore, the campus will be able to operate in islanded mode most of the time without drawing power from the grid. Full islanding capability has also been tested [3, 26].
- The Fort Collins Demonstration Project is a subproject of a larger project called Fort ZED (Fort Collins Zero Energy District), aiming to achieve a 20–30% reduction in peak load on two concentrated distribution feeders in Fort Collins, Colorado. The technologies and programs involved in this project include solar systems, CHP, micro-turbines, fuel cells, plug-in hybrid electric vehicles, thermal storage, demand response, and demand-side management. The peak load is about 46 MW, while the combined capacity of distributed generation and load shedding is only 5 MW [3, 27].

Other research and development projects on microgrids in the United States include the UC San Diego microgrid, Borrego Springs microgrid, and military microgrids at Camp Pendleton and Fort Hunter Liggett.

The UC San Diego (UCSD) microgrid, covering 450 hectares, with two 5.13 MW gas turbines, a 3 MW steam turbine, and a 2.1 MW solar photovoltaic system, meets approximately 85% of the university's power needs. UCSD uses a primary controller called Paladin, responsible for controlling all generators, storage systems, and loads with hourly calculations, and optimizing the microgrid operating conditions. The Paladin utilizes renewable energy forecasting along with dynamic pricing in the market to control renewable energy resources. The solar photovoltaic systems are equipped with advanced controls to achieve optimal performance [3].

Borrego Springs is a small residential community located 90 miles northeast of San Diego. It can be described as an industrial-scale microgrid with active involvement from San Diego Gas and Electric (SDG&E) company, focusing mainly on providing smart grid services. The total installed capacity in this microgrid is approximately 4 MW, including 2 diesel generators (each one is 1.8 mw), a large 500 kW/1500 kWh battery, three smaller batteries less than 50 kWh, six home energy storage units of 4 kW/8 kWh, approximately 700 kW of rooftop solar photovoltaic system, and 125 units of residential network. The main technologies in this microgrid include the FACTS, blackout management system, and price-based load management system [3].

The first implemented commercial microgrid project was the Santa Rita prison microgrid, also known as the "Green Prison Project". This microgrid was developed to demonstrate the reliability of a fully renewable microgrid with energy storage on a large scale. The microgrid includes a 2.1 MW rooftop photovoltaic system, a 1 MW molten carbonate fuel cell with CHP, 2 diesel generators totaling 4.2 MW for emergency situations, 2 kW wind turbines, a 2 MW lithium-ion battery, as well as lighting and HVAC loads. This system has demonstrated dynamic islanding capability with seamless islanding and reconnection. The microgrid is jointly operated by Chevron Energy Solutions, Alameda County, Satcon Power Systems, and Pacific Gas & Electric Company, using financial resources injected from federal, state, and industrial funds [1]. Further details on this microgrid are provided in [3].

7.7.2 Canada

In Canada, the British Columbia Institute of Technology (BCIT) has designed and implemented a 2.1 MW microgrid pilot project at the Burnaby campus in Vancouver, as shown in Fig. 7.7.

This microgrid consists of a campus with two wind turbines (each 5 kW), PV modules (300 kW), a thermal turbine (250 kW), a lithium-ion battery (550 kWh), and campus loads (electric vehicle charging stations, industrial loads, classrooms, offices, and residential areas) [28]. This microgrid is composed of the following subsystems [29]:

- The Open Access to Sustainable Energy Resources (OASIS) subsystem includes a 250 kW PV array, 500 kWh lithium-ion batteries, a 280 kW grid-connected inverter with islanding capabilities, and clusters of fast DC electric vehicle charging stations and other campus area loads.
- The Steam Turbine Generator (STG) subsystem includes a 250 kW steam turbine with 75 kWh lithium-ion batteries and a 25 kW grid-connected inverter. There are removable loads that are powered by this turbine.
- The Gateway Building Microgrid subsystem consists of 84 PV cells with an installed capacity of 17 kW, which powers the second-floor electrical panel of the building and includes an electricity meter that measures the PV generation.

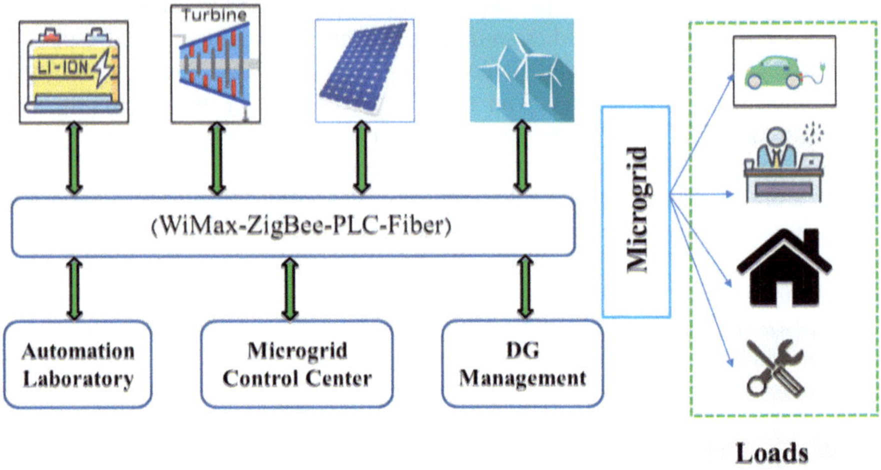

Fig. 7.7 Schematic of the BCIT microgrid

- The Smart Home subsystem is a zero-energy consumption home consisting of a 4 kW PV array, a 5 kW wind turbine, 4 kW lead-acid batteries, an electric vehicle charger, smart appliances, smart meters, internal displays, smart thermostats, heating and cooling systems, and an energy management system. An electricity consumption meter measures and records the home's power consumption.

7.7.3 Europe

In European countries, many research and development projects have focused on microgrids, smart grids, energy efficiency, renewable resources, and distributed generation, often funded by national or EU budgets. For example, in Denmark, Energinet.DK aims to build a complete pilot system (the Eco Grid EU project) with over 50% of its energy generation coming from renewable sources. In Germany, RWE DAG is developing a concept of a virtual power plant, while in Austria, Salzburg AG is constructing a prototype building to evaluate the benefits and drawbacks of smart grids in relation to buildings. Other important projects in Europe include the PREMIO project by Electricité de France (EDF) in the PACA region of France, and the residential microgrid project Am Steinweg in Stutensee, Germany. These projects involve various forms of distributed generation, storage systems, renewable energy sources, and connected loads.

The Bronsbergen Holiday Park in the Netherlands features 108 buildings with photovoltaic modules, making it a sustainable and eco-friendly accommodation option. These solar panels help reduce the park's carbon footprint and reliance on traditional energy sources. The system has a maximum generation capacity of 315 kW and a centralized energy storage system. The park is capable of generating power for 208 accommodation units with a maximum total load of 105 kW. Furthermore, two batteries are present for energy storage and supply when solar power is not available. A switch is used to connect the transformer to the 10 kV medium-voltage grid. Further details about this microgrid are provided in [30].

In Italy, CESI Ricerca DER test bed is mentioned as a low-voltage microgrid (400 V), comprising several DERs and controllable loads that are connected to a medium-voltage grid through an 800 kVA transformer and suitable converter. Additionally, a smart grid project at La Sapienza University in Rome incorporates internal combustion engines, photovoltaic systems, fuel cells, micro gas turbines, and absorption chillers for electricity and heat generation [31].

The infrastructure established at Université de Technologie de Compiègne (UTC) in France is an example of a residential microgrid. Its main local source is a photovoltaic array on the roof integrated with a unified storage system. These devices are connected to a DC bus, which is then linked to the AC section of the grid through a converter. The implemented tertiary control allows for maximizing local consumption from photovoltaic generation. Islanding operation is also feasible as the local generation can supply all local loads, and the control equipment enables voltage and frequency regulation [32].

The microgrid of Suil University in Spain is designed as a small-scale microgrid at the household level and its control is programmable logic controllers (PLC). This microgrid consists of an electrolyzer and a fuel cell generator as an energy storage system. The surplus power from the photovoltaic farm is used by the electrolyzer to produce hydrogen for time when solar power generation is insufficient or unavailable, which is then converted to electrical power by the fuel cell generator. Lead-acid batteries are also used for voltage regulation. All equipment are connected to a main DC bus and have a dedicated DC/DC converter. The main DC bus is then connected to local loads and the distribution grid via an inverter [33].

7.7.4 *Japan*

Several relevant initiatives have been carried out in Japan, mostly by the New Energy and Industrial Technology Development Organization (NEDO). The first project is the Aichi project. This project started at the 2005 World Exhibition and then was moved to central Japan airport near Nagoya in 2006, where it now supplies electricity to an administrative building and a wastewater treatment plant through a private distribution line. This microgrid uses fuel cells, including two molten carbonate fuel cells with a nominal capacity of 270 kW and 300 kW, four phosphoric acid fuel cells of 200 kW, and one 50 kW solid oxide fuel cell as primary sources. The molten carbonate fuel cells are fed by biogas produced from high-temperature (1200 °C) wood and plastic waste gasification. Both molten carbonate and solid oxide fuel cells are used for base load supply, while the phosphoric acid fuel cells are used for regulation. There is also a 330 kW photovoltaic generation and a

500 kW sodium-sulfur (NaS) battery for balancing purposes. This system can operate both connected to the grid and in islanded mode. More details about this microgrid are presented in [3].

Another example of a microgrid project is the Hachinohe microgrid project in Aomori Prefecture, Japan [34]. Its main features are that it provides both electrical and thermal energy, and it relies solely on renewable and biomass sources. The system consists of three 170 kW generators (for a total of 510 kW nominal capacity) fueled by biogas from sewage digestion, a 100 kW lead-acid battery bank, and a biomass boiler fueled by 1 ton/h of wood. There are also photovoltaic farms (2 × 50 kW and 3 × 10 kW) and small wind turbines. The microgrid is capable of supplying energy to seven buildings in Hachinohe City. A 6 kV distribution line, 4.5 km length with dual circuits for redundancy, is the main connection infrastructure. Significant savings have been achieved thanks to this infrastructure. During the initial operation period from November 2005 to July 2006, a 3.57% reduction in primary energy consumption was estimated, while carbon emissions decreased by 8.47%. An islanding test was also conducted for 1 week during this period. The microgrid energy management system, developed specifically for this project, is responsible for optimizing the building's electricity and thermal power needs by controlling the output of generators and boilers, as well as the energy exchanged by storage equipment, with the aim of minimizing operational costs, carbon dioxide emissions, and maintaining a steady power flow at the PCC. This microgrid was developed as a fully renewable microgrid by the regional power grid as part of the budget of the New Energy and Industrial Technology Development Organization. The microgrid was active from October 2005 to March 2008, but became non-operational due to non-financial constraints. This project was carried out in collaboration with the city of Hachinohe, Mitsubishi Research Institute, and Mitsubishi Electric. The aim of this project was to develop and demonstrate the optimal operation and control of a microgrid system to evaluate power quality, operational costs, and emission reductions. More details about this microgrid are presented in [3].

7.8 Conclusion

In this chapter, definitions, concepts, principles, applications, and components of microgrids and nanogrids were discussed. After introducing the microgrid and explaining its concept, definitions related to microgrids and their features were presented. According to the provided definitions, the most important advantages of microgrids include ensuring reliable and flexible power supply, reducing transmission losses, minimizing system capacity utilization, and utilizing renewable energies. In order to benefit from these advantages, stability, control, protection, islanding detection, and security are the most significant challenges that microgrids face. Furthermore, various applications of microgrids were examined. In the next step, the design method of a microgrid and the software used for its simulation were introduced and discussed. The usable software includes HOMER, MATLAB,

PSCAD, GridLAB, and others. IEC and IEEE standards associated with microgrid and nanogrid design and operation were introduced focusing on their applications. Finally, recent projects developed in the USA, Canada, European countries, and Japan were presented. Based on the mentioned, a smart city with microgrids or nanogrids is a sustainable urban environment that incorporates advanced technologies to efficiently manage energy resources.

References

1. S.K. Kottayil, *Smart Microgrids* (CRC Press, Boca Raton, 2021)
2. N. Mahdavi Tabatabaei, E. Kabalci, N. Bizon, *Microgrid Architectures, Control and Protection Methods* (Springer, Cham, 2020)
3. N. Hatziargyriou, *Microgrid: Architectures and Control* (Wiley, Hoboken, 2014)
4. IEC TS 62898-1, Microgrids—Part 1: Guidelines for microgrid projects planning and specification, 2017
5. Asian Development Bank, Handbook on Microgrids for Power Quality and Connectivity, July 2020
6. OPAL-CHART, Website da OPAL-RT, 2019. https://www.opal-rt.com/
7. OPAL-BROCHURE. Real-time simulation solutions for power grids and power electronics, 2019. https://www.opal-rt.com/wpcontent/themes/enfoldopal/pdf/L001610260.pdf
8. TYPHOON, Website da Typhoon HIL, 2019. https://www.typhoon-hil.com/
9. RTDS Technologies, Microgrid simulation and testing. https://www.rtds.com/
10. IEC TS 62898-2, Microgrids - Part 2: Guidelines for operation, 2018
11. IEC TS 62898-3-1, Microgrids - Part 3-1: Technical requirements - Protection and dynamic control, 2020
12. IEEE 2030.7, IEEE standard for the specification of microgrid controllers, 2017
13. IEEE 2030.8, IEEE standard for the testing of microgrid controllers, 2018
14. IEEE 2030.9, IEEE recommended practice for the planning and design of the microgrid, 2019
15. IEEE 2030.10, IEEE standard for DC microgrids for rural and remote electricity access applications, 2021
16. IEEE P2030.12, Guide for the design of microgrid protection systems, https://standards.ieee.org/project/2030_12.html
17. IEEE 1547, IEEE standard for interconnection and interoperability of distributed energy resources with associated electric power systems interfaces, 2018
18. IEEE 1547.1, IEEE standard conformance test procedures for equipment interconnecting distributed energy resources with electric power systems and associated interfaces, 2020
19. IEEE 1547.2, IEEE application guide for IEEE Std 1547(TM), IEEE standard for interconnecting distributed resources with electric power systems, 2008
20. IEEE 1547.3, IEEE guide for cybersecurity of distributed energy resources interconnected with electric power systems, 2023
21. IEEE 1547.4, IEEE guide for design, operation, and integration of distributed resource island systems with electric power systems, 2011
22. IEEE 1547.6, IEEE recommended practice for interconnecting distributed resources with electric power systems distribution secondary networks, 2011
23. IEEE 1547.7, IEEE guide for conducting distribution impact studies for distributed resource interconnection, 2013
24. R.H. Lasseter et al., CERTS microgrid laboratory test bed. IEEE Trans. Power Deliv. **26**(1), 325–332 (2011)
25. N. Hatziargyriou, H. Asano, R. Iravani, C. Marnay, Microgrids. IEEE Power Energy Magaz. **5**(4), 78–94 (2007)

26. A. Flueck, Z. Li, Destination: Perfection. IEEE Power Energy Magaz. **6**(6), 36–47 (2008)
27. M. Panwar et al., Dispatch in microgrids: Lessons from the Fort Collins renewable and distributed systems integration demonstration project. Electr. J. **25**(8), 71–83 (2012)
28. H. Farhangi, *Smart Microgrids: Lessons from Campus Microgrid Design and Implementation* (CRC Press, Boca Raton, 2017)
29. H. Farhangi, G. Joos, *Microgrid Planning and Design: A Concise Guide* (IEEE Press, Piscataway, 2019)
30. R. Bayindir, et al., Microgrid Facility at European Union, ICRERA 2014, Milwaukee, WI, October 19–22, 2014
31. G. Dondossola, et al., A laboratory testbed for the evaluation of cyber attacks to interacting ICT infrastructures of power grid operators, SmartGrids for Distribution IET-CIRED Seminar, 2008
32. M. Sechilariu, B. Wang, F. Locment, Building integrated microgrid: Advanced local energy management for forthcoming smart power grid communication. Energ. Buildings **59**, 236–243 (2013)
33. L. Valverde, C. Bordons, F. Rosa, Power Management Using Model Predictive Control in a Hydrogen-Based Microgrid, in *IECON 2012 - 38th Annual Conference on IEEE Industrial Electronics Society*, (IEEE, Piscataway, 2012)
34. Y. Fujioka et al., *Regional Power Grid with Renewable Energy Resources: A Demonstrative Project in Hachinohe* (CIGRE, Paris, 2006)

Chapter 8
Smart Home and Building Implementation in Smart Cities

Amirreza Samadi Bonab and Parisa Shirinabadi

8.1 Introduction

Electricity is a vital resource that has been essential in people's lives for many years. Today, as energy consumption has increased, power plants have also increased production. Numerous countries have implemented it [1]. Additionally, the advent of smart grids and renewable resources can provide a new perspective for optimizing energy consumption [2].

Ensuring clean, secure, and efficient energy is paramount today. Electricity, a stable energy source, is crucial in laying a robust foundation to address these challenges [3]. Since 1990, the energy demand has increased by almost 40% [4].

To provide a suitable environment for the residents and to use smart homes and smart devices to increase the efficiency of the equipment and the well-being of the residents, suitable infrastructures should also be created [5]. Smart networks are one of the best solutions to control energy consumption. Smart grids refer to smart electric grids that use advanced communication and control technologies. Their key goals include increasing reliability, minimizing energy consumption, incorporating renewable energy sources (RESs), and facilitating effective two-way communication between service providers and consumers [6].

Photovoltaic (PV) systems are the primary electricity generation resource for smart homes .Efficient planning of residential loads can help save on electricity bills and provide opportunities to participate in energy markets for dwellings equipped with energy production sources [7].

The advancement of power systems toward smart grids has enhanced load management, reliability, and the integration of various types of renewable energy sources

A. Samadi Bonab (✉) · P. Shirinabadi (✉)
Faculty of Electrical and Computer Engineering, University of Tabriz, Tabriz, Iran
e-mail: amirrezasamadibonab@ieee.org

A. Ahmadian et al. (eds.), *Sustainable Energy Resources in Smart Cities*, Green Energy and Technology, https://doi.org/10.1007/978-3-031-91516-1_8

(RESs), such as wind and solar units. However, the variability of RESs presents challenges and new obstacles to the efficient management and operation of smart distribution systems (SDSs) [8].

Smart homes have become a topic of interest due to global warming and environmental issues; governments are now considering the use of clean and renewable energy more than ever. In the future, renewables must compete with other energy sources without support from demand-side policies [9].

In this chapter, another important application of IoT in electric cars, which has attracted the attention of many countries today due to its numerous benefits, is discussed. Car manufacturers are striving to make the car manufacturing process more intelligent, environmentally friendly, and self-reliant. Therefore, old cars with outdated features are gradually abandoned, and new cars with updated, advanced, and smart features are entering the market. One of the most essential advantages of these cars is their impact on nature and the environment. These cars are similar in appearance to previous models; nevertheless, pollution is minimized due to the absence of fossil fuels, leading to the conclusion that these cars are more eco-friendly. In these vehicles, a system called wireless sensors allows users to monitor parameters such as battery efficiency, charging alerts, and distance traveled. With these sensors, the journey becomes straightforward and safe for drivers, as they can detect heavy traffic and manage conditions. The cost would be significantly higher if these features were installed in standard and older cars. Additionally, the integration of smart homes and electric cars enables users to control and check the charging of the car and status from anywhere [10].

Another critical aspect of IoT is load response ability. In this technique, demand is managed, and energy is consumed in the most optimal way; the load is reduced during peak times and increased during off-peak times. This type of management called demand response (DR) can be used in many sectors, such as commercial, industrial, or residential. In smart homes, sensors detect the amount of load and send that data to a local controller. This way load consumption in that area is monitored, and, during peak times, consumption is reduced to be more economical [11].

Another critical option offered by smart homes is the heating and cooling system. This system allows people to control the temperature as they see fit, considering certain limits and efficiency. When homeowners are not at home, high temperatures may waste energy. Heat production is associated with energy consumption, and when the temperature balance is not controlled, more energy is wasted, which is indeed a costly decision.

Due to these problems, smart homes have created a suitable solution for people. In the following chapter, each of the topics above will be further discussed.

8.2 Smart Home

8.2.1 The Effects of Smart Homes on Energy Management

One of the most important benefits of integrating smart homes and smart cities is optimal energy management. Smart homes can communicate with smart networks and adjust energy consumption based on real needs and programmed prices. This not only helps reduce energy costs but also leads to the stability of the power grid and the reduction of peak loads.

Homes that have become smart, equipped with controllers, sensors, and advanced systems, can significantly reduce energy consumption and manage energy levels during peak times.

By making homes smart, energy consumption is reduced by an average of 30% and peak demand by 20% [12]. This reduction will decrease expenditure, and this method can increase productivity. Therefore, in the future, we can significantly help the environment. Traditional and old houses that used common appliances and lacked energy management systems were often inefficient and not smart. Due to this issue, the concept of smart homes has received considerable attention to enhance energy efficiency in homes and reduce costs.

In networked smart homes, a set of devices such as automatic electric curtains, LED lighting systems, smart thermostats, and other devices communicate with each other using mobile phones and advanced internet or a series of commands and codes. To perform remote control, these intelligent systems can automatically execute energy-saving commands according to the instructions and codes they receive from the commander.

8.2.1.1 Renewable Energy Source (RES)

Global warming has become a significant challenge and concern in recent times. Energy production companies view the use of renewable energy sources (RESs) as an appropriate solution to address global warming and reduce carbon emissions [13].

Considering that nowadays, with the increase in the number of pollutants and carbon gas, the country's and people's need to use clean energy has increased, people can reduce their energy consumption by combining renewable energies such as wind turbines or solar energy with this technology. Control and use clean energy. When people breathe this gas, it has adverse effects on their health.

Therefore, despite these problems, using clean and renewable energies such as solar and wind power will be the best solution.

These energies use clean and accessible resources from the environment and produce energy (Fig. 8.1).

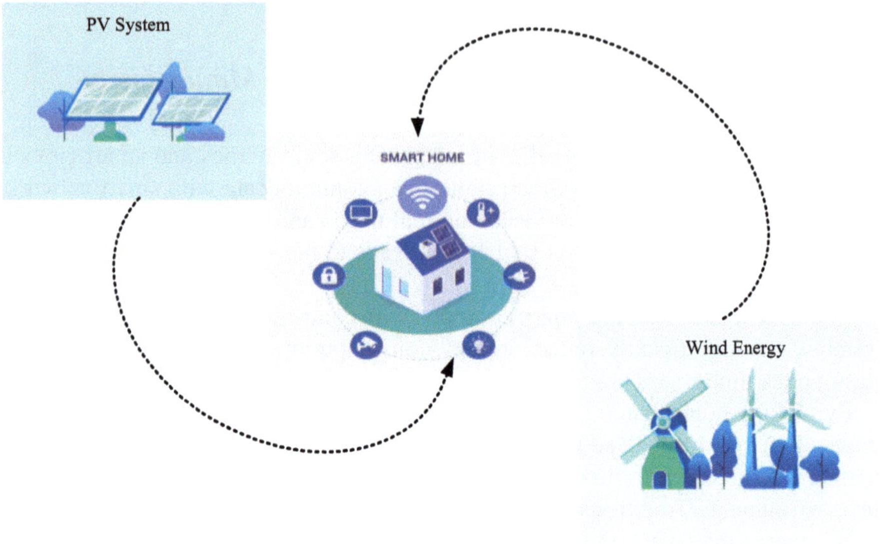

Fig. 8.1 Renewable energy sources for smart homes

8.2.2 Smart Power Outlet

Global warming has become a significant challenge and concern in recent time. Energy production companies consider using renewable energy sources (RESs) as the appropriate solution to address global warming and reduce carbon emissions [13].

Considering that nowadays, with the increase in the number of pollutants and carbon emissions, the country's and people's need to use clean energy has grown, people can reduce their energy consumption by integrating renewable energies such as wind turbines or solar energy with this technology. This allows for better control and utilization of clean energy. When people breathe in these pollutants, it has adverse effects on their health.

If the smartization of cities by incorporating renewable energy sources becomes favorable, many environmental problems will be solved because carbon dioxide production will decrease.

People can remotely control home electrical appliances remotely by connecting to Wi-Fi or Bluetooth.

Smart sockets are very useful in reducing energy consumption. So that some devices consume energy when they are turned off and want to be started. By using smart sockets, these types of devices can be completely turned off and energy wastage will be prevented.

The smart switch and socket are fantastic products that have attracted attention. The following are the benefits of this product (Fig. 8.2):

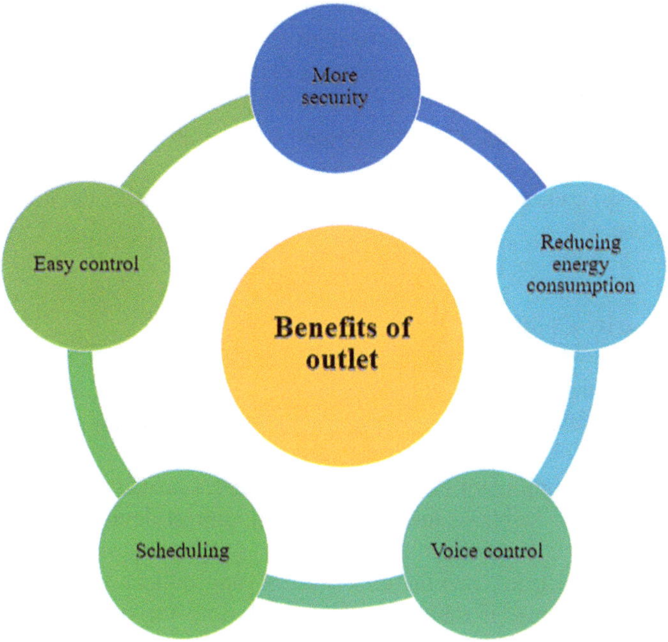

Fig. 8.2 Benefits of the smart switch and socket

In short, smart switches and sockets allow people to control their home electrical appliances intelligently and remotely, which can save energy and improve user comfort.

Smart sockets contribute to energy efficiency in several ways. First, they enable users to remotely turn appliances on or off, preventing unnecessary energy consumption when they are not in use.

Additionally, they provide real-time power usage data, allowing users to identify devices with high energy consumption and consequently reduce wasted electricity by changing their behavior.

When integrated into smart home automation systems, these plugs can cooperate with other devices, such as intelligent lighting or zoned heating, to create comprehensive energy-saving strategies.

8.2.3 Safety and Security

In the past, smart homes had a more straightforward concept called protection; however these days, with the advent of smart homes and the advancement of technology, its meaning and concept have become a little more complex.

In other words, with the development of smart homes, the issue of security is taken more seriously, and efforts are made to enhance security.

In this section, the concept of security is explained; then, we discuss the challenges and proposed solutions.

Security in smart homes refers to specific technologies and measures that provide suitable conditions to protect property and valuables, homes, and even human lives. Some smart cameras, sensors, warning systems, and other devices are utilized for this purpose.

8.2.3.1 Security Challenges

Smart homes improve people's lives and increase their security by providing various facilities. However, despite these advantages, there are many challenges associated with them.

One of the most significant security issues in smart homes is cyber-attacks. If the home network is accessible to hackers, they can control devices connected to the Internet and steal personal information.

Smart devices such as cameras and sensors may collect sensitive and private information of residents. If this information falls into the hands of a hacker, the privacy about residents will be severely threatened.

Smart homes can be integrated with city security systems. Security cameras and sensors installed in homes can send data to city monitoring centers, providing a faster response by rescue forces in incidents such as theft or fire. This cooperation can lead to increased security and peace of mind for residents [14]. Many of these homes use Wi-Fi networks to connect devices. If these networks are not functioning correctly, hackers can quickly access them.

Phishing attacks are another security challenge in smart homes, as hackers send suspicious emails and messages to residents to steal their important information [15].

8.3 Key Technologies

Some technologies, such as Blockchain, artificial intelligence, and the Internet of Things, are used in smart homes to make daily life easier.

8.3.1 Internet of Things (IoT)

The concept of smart homes has become a controversial issue that can be examined from different aspects. These homes have gained attention for various reasons; one of the most important is that they make the lives of residents much more manageable.

These houses include some technologies and tools that household equipment, in utilizing data analysis to predict behavior and finally control it [16].

The security of a house can be increased by using these systems. With some technologies that communicate with mobile phones and connect to them, and with an Android application on the mobile phone, home security can be controlled and monitored.

The Internet of Things enables the connection between various physical devices embedded with sensors, software, and other technologies that make this networking possible. The ultimate goal is to achieve an automated environment through data exchange over the Internet.

Nowadays, the risk of home theft has increased. Therefore, the issue of security has become a serious concern for everyone. To prevent the risk of property theft, a security system should be used, allowing people to control the security of their homes using a mobile phone or laptop.

8.3.2 Artificial Intelligence (AI)

Artificial intelligence has been one of the most significant achievements of humanity, transforming the complex and laborious life of humans into a comfortable and less effortful existence. Despite this technology and essential achievement, people can accomplish many tasks at home without any special effort, simply by pressing a button on their mobile phones.

Smart homes, which are the main subject of this book, utilize this same artificial intelligence technology. In fact, this entire process depends on the existence of artificial intelligence technology. People with a smart mobile phone can intelligently control and manage many of their household chores.

Artificial intelligence is a system in which a human gives an order by pressing a button, and the artificial intelligence receives the order, analyzes it, and executes it, thereby improving the quality of human life [17].

8.3.3 Blockchain

Blockchain is a special technology that plays an important role in many areas, such as security. Naturally, data security and privacy are essential for users. Blockchain technology allows users to record any type of transaction or data exchange somewhere securely, and this is a great way to protect data and protect against theft.

Blockchain also plays a vital role in energy management and control. By using this technology, individuals can control the amount of energy consumption in their homes to prevent wastage and utilize energy in its most optimal state.

Another application of Blockchain is in resource sharing. Imagine people within a building, each living separate lives, with varying energy consumption in their homes. In this scenario, Blockchain enables people to share resources, such as renewable energy or Internet access.

8.4 Smart Homes in Smart Grid Ecosystem

Integrating smart homes into the smart grid ecosystem signifies a pivotal stride in proficient and astute energy management. Smart homes, driven by interconnected devices such as smart meters, thermostats, energy storage systems (ESSs), and communication interfaces with the grid, contribute to maintaining an equilibrium between energy supply and demand. Real-time access enables the implementation of dynamic pricing and demand response programs, leading to decreased electricity expenditure for consumers. Moreover, this integration fortifies grid stability by effectively redistributing surplus energy generated by renewable energy sources (RESs) and accumulating it for future use, propelling us toward a more sustainable future.

8.4.1 Smart Grid

A smart grid is a contemporary concept of electrical power grids that enables bi-directional communication between the utility provider and the consumer. It allows for the integration of Distributed Energy Resources (DERs) and access to real-time data through the implementation of the Internet of Things (IoT), sensors, and smart meters. This leads to sustainability, efficient energy and cost management, reliability, and accurate monitoring, allowing for real-time tracking of consumption and production, as well as controllability by the Grid Operator [18] (Fig. 8.3).

8.4.2 Enabling Technologies for Smart Home Integration with the Grid

The vision of a more sustainable future is only possible if the integration is feasible, which can only be achieved through some gadgets. Simply put, advanced technology is required to enable bi-directional communication, real-time data processing, and intelligent control mechanisms to optimize energy consumption and distribution, enhance grid stability, and facilitate user responsibility.

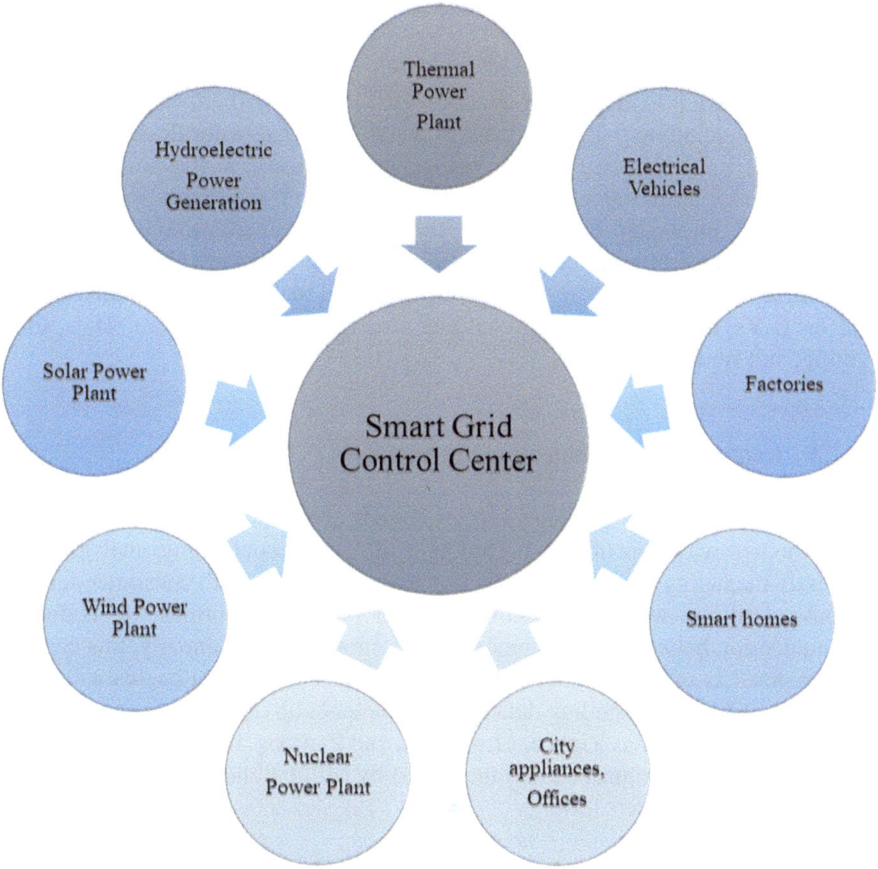

Fig. 8.3 Smart grid system overview

8.4.2.1 Smart Meters

Smart meters are digital devices that measure, record, and communicate high-resolution electricity usage data frequently, ranging from a few seconds to a few minutes. This data is transmitted to the utility provider, the customer, or home energy management systems (HEMSs) in smart home for monitoring and billing purposes. They are capable of real-time consumption monitoring, supporting time-of-use pricing models, outage detection, and remote reading services.

It can measure voltage by manipulating a resistive or capacitive divider and current by employing a current transformer (CT) or Hall effect sensors. Based on the acquired data, it calculates apparent power, active power, and reactive power.

8.4.2.2 Home Energy Management System (HEMS)

Smart meters are usually connected to HEMS. The data collected by the Smart Meter is transferred to HEMS for analysis. HEMS exploits hardware and software technologies, including IoT devices, smart meters, sensors, controllers, and machine learning (ML) algorithms to enable real-time monitoring, analysis, and management of energy transactions in households [19]. IoT facilitates real-time data exchange and automated coordination. To be able to process the vast amount of information induced by various residential devices and provided by the grid. This data comes from DERs, namely PV roof panels and their storage systems, electric vehicles (EVs) and their charging facilities, smart meters, smart thermostats, smart lighting, and smart plugs.

8.4.2.3 Required Communication Technologies

The smart grid requires a robust and intelligent communication infrastructure to enable the two-way flow of massive amounts of data for controlling, analyzing, and efficiently managing electricity production and distribution with consumer demand. Various wired and wireless technologies offer potential solutions for facilitating communication between smart homes and the smart grid. Essentially, this communication consists of two distinct steps: The transmission of data begins with home appliances and sensors sending data to smart meters utilizing wireless communication technologies such as Zigbee, 6LowPAN, and Z-wave. The subsequent transfer of data to the utility control center, facilitated by cellular and internet connections, represents valuable assets [20].

8.4.3 Smart Homes Asset for the Smart Grids

Smart buildings play a significant role in the automated grid, as they account for a substantial share of electricity consumption. In the USA, it has been estimated that this share is around 40%, as demonstrated in Fig. 8.4, where the consumption of commercial and residential buildings adds up to 40%. Energy consumption management will be possible since smart buildings can charge and discharge even in groups (multi-agent systems) [21]. The integration of smart homes in smart grids facilitates two-way communication and energy exchange between them, which is made possible by smart meters, HEMS, energy storage systems (ESS), and more.

Smart homes have a few boons regarding smart grids:

- Demand response and load shift.
- Integration of Distributed Energy Resources (DERs).
- Higher energy efficiency.
- Enhanced grid reliability and resilience.

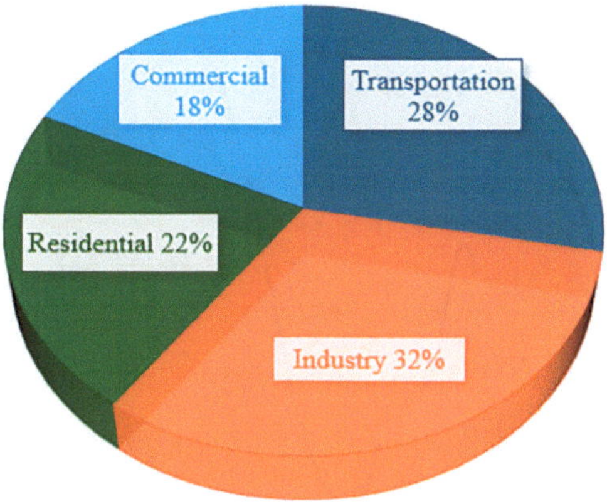

Fig. 8.4 US energy use pie figure

These benefits will be discussed further in the following sections.

8.4.3.1 Demand Response and Load Shift

Demand response (DR) and load shifting both aim to optimize energy usage by determining the level and timing of consumption. The significant outcome is not only to balance supply and demand (reducing peak time load), which leads to enhanced grid stability but also to make better use of available resources in a cost-effective way. They both rely on automation and intelligent data accessibility, provided by technologies such as smart meters, smart thermostats, HEMS, ESS, and other smart appliances. Nevertheless, this is all done under the supervision of the house owner, who is enticed to participate in these programs' financial rewards.

Demand response involves reducing and shifting energy consumption triggered by real-time grid signals. The action should be immediate and require more precise and complex control systems. It is primarily a response to real-time grid conditions, such as critical peak periods that expose the power system to stress or shifting usage based on real-time price signals, however temporary. It can be viewed through in two ways:

- Incentive-based DR: This approach includes remote load control by the utility and manual load reduction for payment or bill exchange.
- Time-based DR: Consumers are encouraged to take advantage of off-peak hours when the electricity is low-priced.

On the other hand, load shifting is a less urgent scheduled route. It is based on knowledge of electricity's dynamic pricing and the owner's decision to exploit this information to reduce the bill (Fig. 8.5). Simply put, it shifts energy-exhaustive activities to off-peak times rather than reducing consumption. It is scheduled based on the customer's predictable, less congested grid times and does not require high-tech devices; it can be managed by a basic timer and smart appliances [22].

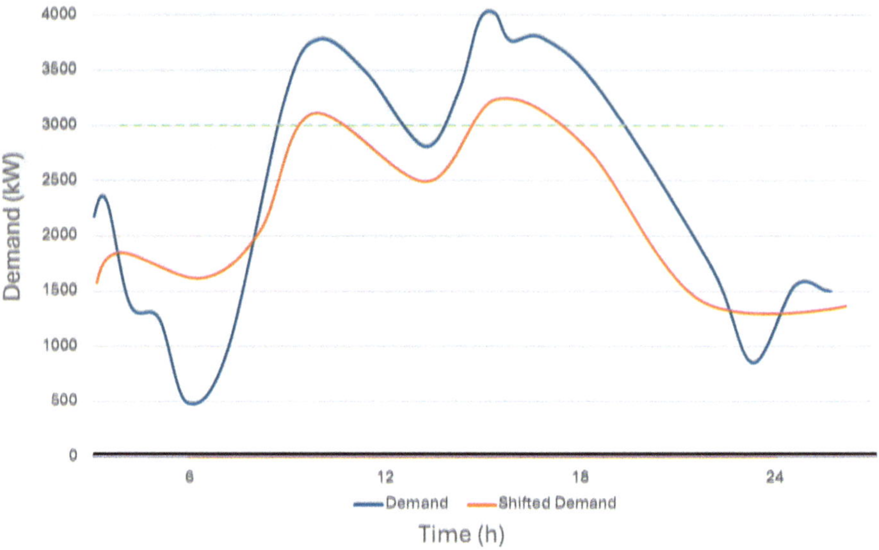

Fig. 8.5 Load shifting

As mentioned, smart homes rely on real-time data, such as signals from the electrical grid, that makes automatic load shifts possible.

8.4.3.2 Integration of Distributed Energy Resources (DERs)

Distributed Energy Resources (DERs) are decentralized units that generate, store, and manage electricity on a small scale, often situated adjacent to the point of consumption. They consist of micro-turbines, diesel or natural gas generators, solar arrays, battery energy storage systems (ESSs), etc. It is common for them to be located adjacent to the load. They can be used individually or in aggregation, working in synergy with the centralized grid.

Considering the topic, the DERs for smart homes can include rooftop solar panels, battery storage, electric vehicles (EVs) and their associated charging infrastructure, smart thermostats, hot water systems, air conditioning (AC), and other controllable household appliances. Customers can control these assets with assistance from home energy management systems (HEMSs).

HEMS monitors PV panels' real-time energy production and consumption by the devices connected to them to devise a lucrative and environmentally friendly course of action for the owner. It prioritizes the utilization of solar energy over-relying on the grid. For instance, it schedules the EV charging during high solar power generation periods [23].

In short, DER integration in Smart Homes can be an asset since it brings along autonomous green energy production, leading to reduced reliance on the grid. By interfering with HEMS, energy usage is optimized, thus cutting bills. DERs, in

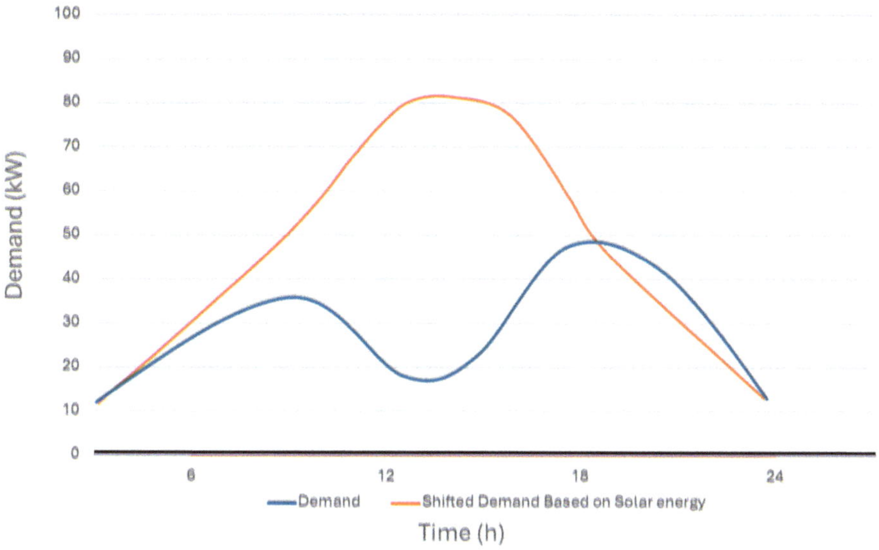

Fig. 8.6 Load shifting based on solar energy availability

correlation with HEMS, can contribute to grid stability by providing ancillary service that promotes maintaining the frequency and voltage levels within permitted limits (Fig. 8.6).

This is a new emerging topic that needs further investigation, and new optimization methods that are practical at the same time need to be employed.

8.4.3.3 Higher Energy Efficiency

Smart homes have interconnected devices that facilitate remote control based on real-time data. The smart meter gathers data on energy usage from various appliances, allowing homeowners to monitor and analyze this information at their discretion. Furthermore, energy storage systems in smart homes can be charged by renewable energy sources such as PV panels on the roof, enabling customers to rely on their generation during peak times and charge during off-peak hours [24]. Integrating of energy storage systems (ESSs) into smart grids optimizes energy efficiency while maintaining grid balance. This is achieved by distributing excess energy during peak production hours and ensuring a steady supply during low production hours with the help of renewable energy sources (RES). Both smart grids and smart homes can forecast demand and learn from past demand patterns, leading to more cost-efficient and optimized operations. Machine learning algorithms play a crucial role in this process by identifying consumption patterns that help minimize costs while considering the consumer's needs. For example, recognizing a pattern of high energy consumption in the evening can help adjust heating and cooling systems accordingly.

8.4.3.4 Enhanced Grid Reliability and Resilience

Smart homes are equipped with sophisticated energy management systems that can autonomously adjust the operation of connected devices by utilizing data from devices, real-time energy pricing, weather forecasts, and grid status to optimize consumption patterns. The Advanced Metering Infrastructure (AMI) facilitates the monitoring and control process for the grid, offering real-time data on energy consumption and enabling swift responses to anomalies. It also identifies incidents such as energy theft, decreased efficiency, and equipment failure, thereby enhancing grid reliability.

Furthermore, the Internet of Things (IoT) creates a collaborative network among smart homes and the grid, optimizing energy usage at a community level. For instance, surplus solar production in one household can be shared with neighboring households, maximizing the utilization of renewable energy sources and reducing demand on the grid.

Incentive-based programs can incentivize homeowners to reduce or shift their electricity consumption to off-peak periods. For example, a smart water heater can postpone heating water until off-peak hours. The cumulative effect of smart homes‘participation in such programs results in decreased power plant production during high-demand periods, which is less efficient than base power production.

Integrated with the smart grid, smart homes significantly balance demand and supply while considering the homeowner's benefits. They enable load shifting and demand response, mitigating pressure during high-demand hours by reducing consumption or delaying the use of appliances to off-peak hours.

In addition to load shifting, smart homes can provide ancillary services to the grid, such as frequency control and voltage regulation. Home batteries and electric vehicles (EVs) can store excess energy when supply exceeds demand and distribute it back when there is a shortfall. This bidirectional energy flow contributes to a more resilient grid that is robust against disruptions. Furthermore, a network of decentralized energy resources in smart homes, such as solar panels, home batteries, and EVs, can operate even if a part of the grid is down, allowing these homes to switch to islanded mode, ensuring uninterrupted supply to consumers, and reducing dependence on the grid, thereby enhancing supply security.

8.5 Conclusion

8.5.1 Smart Cities

At the end of this chapter, we take a brief look at the smart city review. Smart cities, with the help of the government and utilizing intelligence and mobile phone technologies, improve the quality of citizens' lives and help them have an easier life.

In fact, the main goal of smart cities is to help economic growth and improve people's quality of life.

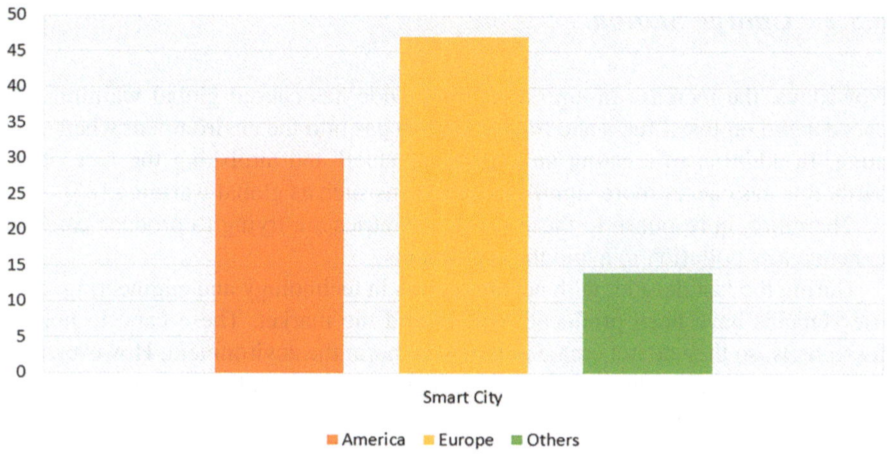

Fig. 8.7 Worldwide distribution of smart city

Smart cities have made significant progress over time, which is more visible in some countries and is a sign of a modern and advanced life. The graph below shows the progress of smart cities in America, Europe, and other parts of the world [18].

According to this graph, the rate of progress of smart cities in Europe is higher than in other parts of the world (Fig. 8.7).

Urban smartening has different aspects, which will be briefly explained below:

1. For example, in smart cities, much attention is paid to environmental issues, and infrastructure is created to prevent environmental pollution and to make public transportation in the city more manageable and efficient.
2. The ITS transportation system is a collection that helps to improve the performance of public transportation of technologies in the city.
3. Smartening the economy in the city is another application of smart cities, in this way, by using ICT, the production of goods improves and businesses in the town prosper.
4. Smart living facilities are one of the most critical applications of smart cities because they are related to life security and educational resources.

In fact, in a smart city, all the daily activities of people are made smart, such as agriculture, education, transportation, etc.

For example, smart agriculture is one of the most critical applications of smart buildings because it determines people's health, hygiene, and physical security. In smart agriculture, instead of using old methods of irrigation and spraying, new and smart methods are used. In this approach, the use of resources is controlled and managed through a mobile phone. In addition to the fact that enough water reaches the products, the number of resources used is also managed and its wastage is prevented.

8.5.2 Charge Station

Nowadays, the increase in emissions worldwide has caused global warming. All cars depend on fossil fuels and release carbon gas into the environment when operating. In addition to creating an unpleasant smell and tarnishing the face of the earth, this gas causes more significant problems such as global warming [25].

Therefore, in response to these issues, countries are trying to produce cars that generate less pollution to avoid these problems.

During the last decade, with advancements in technology and engineering, electric vehicles have been produced and entered the market. These cars do not use fossil fuels, so they do not create pollution or harm the environment. However, with the increase in the production of these cars, a significant challenge facing engineers is the availability of charging stations. For electric cars to become popular in the cities, suitable charging stations must be established to allow for continuous charging which poses a challenge due to the time required to charge these vehicles. Today, scientists are working to manage the charging process of these cars by developing intelligent systems that optimize charging time, waiting time, and the approximate cost of charging. This approach greatly aids helps a lot in the development and expansion of electric cars in urban areas.

8.5.3 Electric Trains

In this section, we will briefly review electric trains. Electric trains are another new technology that has attracted everyone's attention today, and to create it, electric railways are needed.

In the world, the first practical railway was launched by a German engineer named Werner von Siemens, marking a significant start for the creation and expansion of electric trains. As mentioned in the discussion of electric cars, the increase in cars and public vehicles has led to a sharp rise in fossil fuel consumption, which is a concern for the entire world. Therefore, to reduce these pollutants, electric trains have also been developed alongside electric cars. In fact, one of the very important goals of electric trains is to reduce greenhouse gases and provide fast transportation. These trains have a traction system that determines how to obtain power and achieve driving force [26].

This traction system has two types: AC (alternating current) and DC direct current). Both of these systems power the train's electric motors.

These trains reduce the carbon footprint, thus preventing environmental pollution, as they utilize renewable sources such as wind, solar energy, and water.

Therefore, according to the issues that have been reviewed, the discussion of smartization is a very broad topic. Due to the expansiveness of the term, it has attracted everyone's attention, as it is a complex issue that can be investigated from various angles, whether within the city or inside the home. As discussed, the term "smart" is used for both urban facilities and the amenities within each person's home. However, the purpose of creating all these facilities is to improve human life.

Smart homes and smart cities have been developed and received so much attention that all countries are striving to advance human life in this direction. This initiative encompasses all aspects of life, including education, culture, business, agriculture, and household facilities such as lighting, heating and cooling, garden irrigation, house cleaning, etc. However, as we have certain, creating these conditions is not always easy and entails some problems and difficulties.

For example, in the matter of energy management, clean energies such as solar power are utilized; however, not all regions of the country are suitable for use. Naturally, when the energy received from the sun is insufficient, the required electrical energy cannot be created. Therefore, conditions should be provided with appropriate designs to provide the necessary energy in the most optimal way and at the lowest cost.

Another topic that was discussed was electric cars, which are very practical. Additionally, because these cars do not use fossil fuels, they help reduce carbon emissions. Despite the many advantages of electric vehicles, there is a big significant challenge. In their way, and that was the big problem with charging stations.

If electric vehicles are going to increase in the city and people are going to use electric vehicles instead of regular cars, a suitable platform should be provided for them so that everyone can charge their car whenever they need (Fig. 8.8).

Therefore, regarding this issue, charging stations should be established in all locations at suitable distances, and the waiting time for each car to be charged should be adequately controlled and managed with appropriate intelligent systems. Thus, the creation of smart cities or smart homes is a highly complex endeavor that faces many challenges, its way, but today, it has garnered everyone's attention due to its advantages.

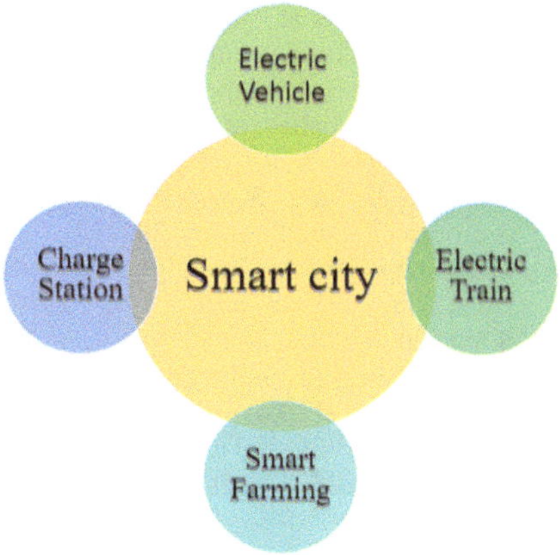

Fig. 8.8 Some parts that we can apply for a smart city

References

1. M. Ghahramani et al., Energy management of electric vehicles parking in a power distribution network using robust optimization method. J. Energy Manage. Technol. **2**(3), 22–30 (2018)
2. H. Jahangir et al., Deep learning-based forecasting approach in smart grids with microclustering and bidirectional LSTM network. IEEE Trans. Ind. Electron. **68**(9), 8298–8309 (2020)
3. A. Almusaed, I. Yitmen, A. Almssad, Enhancing smart home design with AI models: A case study of living spaces implementation review. Energies **16**(6), 2636 (2023)
4. Z.B. Erdem, The contribution of renewable resources in meeting Turkey's energy-related challenges. Renew. Sust. Energ. Rev. **14**(9), 2710–2722 (2010)
5. R. Apanavičienė, M.M.N. Shahrabani, Key factors affecting smart building integration into smart city: Technological aspects. Smart Cities **6**(4), 1832–1857 (2023)
6. M. Keshavarz, A.S. Bonab, *Exploring Innovations in Electrical Engineering* (Nobel Sciences, Pune, 2024)
7. S. Zeynali et al., Two-stage stochastic home energy management strategy considering electric vehicle and battery energy storage system: An ANN-based scenario generation methodology. Sustain. Energy Technol. Assess. **39**, 100722 (2020)
8. H. Jahangir et al., Short-term wind speed forecasting framework based on stacked denoising auto-encoders with rough ANN. Sustain. Energy Technol. Assess. **38**, 100601 (2020)
9. E. Ahmadi et al., Stochastic operation of a solar-powered smart home: Capturing thermal load uncertainties. Sustain. For. **12**(12), 5089 (2020)
10. S. Urooj et al., IoT based electric vehicle application using boosting algorithm for smart cities. Energies **14**(4), 1072 (2021)
11. M.Z. Oskouei et al., A novel economic structure to improve the energy label in smart residential buildings under energy efficiency programs. J. Clean. Prod. **260**, 121059 (2020)
12. R.K. Karduri, A. Gudhenia, The impact of smart homes on energy conservation and demand management. Int. J. Adv. Res. Manage. Arch. Technol. Eng. **4**, 11 (2018)
13. H. Youssef, S. Kamel, M.H. Hassan, Smart home energy management and power trading optimization using an enhanced manta ray foraging optimization. Sci. Rep. **13**(1), 22163 (2023)
14. M. Simić et al., Big data and development of Smart City: System architecture and practical public safety example. SJEE **17**(3), 337–355 (2020)
15. I. Cvitić et al., An overview of smart home iot trends and related cybersecurity challenges. Mobile Netw. Appl. **28**(4), 1334–1348 (2023)
16. O. Taiwo, A.E. Ezugwu, Internet of things-based intelligent smart home control system. Secur. Commun. Netw. **1**, 9928254 (2021)
17. S. Sepasgozar et al., A systematic content review of artificial intelligence and the internet of things applications in smart home. Appl. Sci. **10**(9), 3074 (2020)
18. E.F. Orumwense, K. Abo-Al-Ez, Internet of things for smart energy systems: A review on its applications, challenges and future trends. AIMS Electron. Electr. Eng. **7**(1), 50–74 (2023)
19. G. Prettico et al., *Clean Energy Technology Observatory: Smart Grids in the European Union–2022 Status Report on Technology Development, Trends, Value Chains and Markets* (European Union, Brussels, 2022)
20. V.C. Gungor et al., Smart grid technologies: Communication technologies and standards. IEEE Trans. Industr. Inform. **7**(4), 529–539 (2011)
21. S.S. Ghazimirsaeid et al., Multi-agent-based energy management of multiple grid-connected green buildings. J. Build. Eng. **74**, 106866 (2023)
22. A. Shewale et al., An overview of demand response in smart grid and optimization techniques for efficient residential appliance scheduling problem. Energies **13**(16), 4266 (2020)
23. R. Banihabib, M. Assadi, The role of micro gas turbines in energy transition. Energies **15**(21), 8084 (2022)

24. H. Rehan, Internet of things (IoT) in smart cities: Enhancing urban living through technology. J. Eng. Technol. **5**(1), 1–16 (2023)
25. B. Vaidya, H.T. Mouftah, Smart electric vehicle charging management for smart cities. IET Smart Cities **2**(1), 4–13 (2020)
26. A. Karki et al., Status of pure electric vehicle power train technology and future prospects. Appl. Syst. Innov. **3**(3), 35 (2020)

Index

C
Combined heat and power (CHP), 11, 34–37, 58, 132–135, 139, 142, 143, 177, 196, 197
Controlling, 196, 200, 212

D
Demand response program (DRP), 153, 154, 159–161, 163–166, 168, 169, 171, 172

E
Electric vehicles (EVs), 16, 17, 22, 29, 30, 33, 66, 81, 84, 103, 104, 106, 113, 119, 123, 130, 152, 175, 196–198, 212, 214, 216, 218, 219
Electricity distribution networks (EDNs), 103, 105, 108–109, 111–115, 117–120, 123
Energy consumption, 3, 12, 18, 22, 24, 65, 66, 70, 72, 74, 77, 151, 180, 200, 203–207, 209, 210, 213, 215, 216
Energy consumption management, 212
Energy efficiency, 2, 3, 12–28, 33, 39, 65–92, 132, 198, 205, 207, 212, 215
Energy management, 102, 103, 130–132, 153, 176, 178, 196, 198, 200, 205–206, 209–212, 214, 216, 219
Energy performance certificate (EPC), 68, 71–74, 76, 83, 185, 188
Energy performance of buildings directive (EPBD), 66–71, 74, 75, 78, 90, 92
Environmental impacts, 1, 10, 27, 38, 39, 45, 47, 54, 58, 151, 185

G
Gas distribution networks, 100–124
Gas storage system (GSS), 106, 109, 113, 114, 116–118, 120, 121, 123
Geothermal assessment, 54
Geothermal energy, 1, 5, 8, 45–60
Geothermal potential in Jordan, 47, 54

H
Hybrid energy systems (HSSs), 151–172
Hydrogen fuel cells, 31

M
Microgrids, 34, 102, 129–147, 175–201
Multi-objective optimization, 152, 154, 155, 163

N
Nanogrids, 176–201

P
Peer-to-peer (P2P) transactive energy, 132
Photovoltaic (PV) solar, 152, 161, 167

A. Ahmadian et al. (eds.), *Sustainable Energy Resources in Smart Cities*, Green Energy and Technology, https://doi.org/10.1007/978-3-031-91516-1

R
Renewable energy, 2–4, 30, 34, 45–51, 58, 66, 100–104, 120, 123, 129, 132, 151–153, 180, 186, 189, 194, 196, 200, 204–206, 210
Renewable energy sources (RESs), 2–12, 28, 32, 34, 39, 45, 47, 48, 50, 60, 65, 74, 112, 121, 151, 152, 175, 179, 198, 203–206, 210, 215, 216

S
Smart cities, 1–39, 65–92, 100–124, 146, 175–201, 203–219
Smart energy management tools, 70
Smart energy performance assessment platform (SEPAP), 68–71
Smart grids, 15–18, 27, 39, 175, 189, 190, 197–199, 203, 210–216
Smart homes, 65, 74, 198, 203–219
Smart readiness indicator (SRI), 65–92
Stored heat, 53
Sustainability, 3, 6, 12, 19, 24–27, 36, 37, 65, 92, 152, 185, 186, 210
Sustainable energy, 1–39, 45, 65, 197

U
Uncertainty, 102–104, 110, 111, 120–123, 132, 153
Urban sustainability, 3, 27
User-centric data, 68

MIX
Papier aus verantwortungsvollen Quellen
Paper from responsible sources
FSC® C105338

FSC
www.fsc.org

If you have any concerns about our products,
you can contact us on
ProductSafety@springernature.com

In case Publisher is established outside the EU,
the EU authorized representative is:
Springer Nature Customer Service Center GmbH
Europaplatz 3, 69115 Heidelberg, Germany

Printed by Libri Plureos GmbH
in Hamburg, Germany